AF864879

The Mars-Earth Wars

(Ending 701 B.C.E.)

By Donald W. Patten & Samuel R. Windsor

Reprinted by permission

Copyright 1996, 2003

ISBN: 978-1-4951-1090-0

Published by End Times Thunder Publishers, PO Box 10998, Bozeman, MT 59719
Phone: 1-406-586-4840
Direct inquiries and/or orders to the above address or phone number.

All rights reserved. Except for use in review, no portion of this book may be reproduced in any form without the express written permission of the publisher. Neither the author nor the publisher assumes any responsibility for the use or misuse of information contained in this book.

Printed January 2020

CONTENTS:

Dedication & Credits

This volume is dedicated to Johannes Kepler (1571 - 1630). Kepler is the father of celestial mechanics and the discoverer of three primary laws of planetary motion. He was a friend of Galileo and he was the mentor of both Edmund Halley of cometary fame, and Isaac Newton of Principia fame. The works by Kepler and Newton have been the two biggest birth pangs in the birthing of modern science.

Consultants: Ronald R. Hatch, physicist and orbital analyst
Michael T. Lute, computer technician

William S. Thompson III, astrogeophysicist

Appreciation is expressed to NASA and the Jet Propulsion Laboratory, California, for allowing the use of the reproduction of plates for the following figures: 5, 7, 10 and 11.

Preface

"The Torn Universe and the Destruction of Rahab, Lucifers Planet" - by Stephen Quayle

The pillars of heaven tremble and are astonished at his reproof. He divideth the sea with his power, and by his understanding he smiteth through the proud.

Job 26:11-12

The kingdom was virtually a ghost town. Those that could not get off world had locked themselves behind closed doors, waiting for the inevitable. He was coming for them.

They had made a life here, a good life. They were the makers of their own future and they did what they wanted with "their planet". Why should they be subject to the will of the Creator? After all, now they were their own creators. They had proved they were worthy of this power. The evidenced existed all round them in the numerous hybrid animals and beings they had created in with their technology or enhanced. He had no right.

Out of the east, the whispered whistled caused them to huddle together as they resolved their fate, taking strength in numbers. The whistle turned to a roar and the sky on the horizon blossomed red as the planet's oxygen caught fire.

Suddenly, the ground jolted and a hollow thump was all that was heard before a brilliant white light flashed and pulverized the planet.

Rahab was no more.

In the Beginning...

"In the beginning God created the heaven and the earth. And the earth was without form, and void; and darkness was upon the face of the deep. And the Spirit of God moved upon the face of the waters." Genesis 1:1-2

Take a moment to consider what George H. Pember says about this passage, as it is pertinent to the next part of our study:

> [1]*According to our version, "the earth was without form, and void." This, however, is not the sense of the Hebrew, but a glaring illustration of the influence of the chaos-legend. Fuerst gives "ruin," or "desolation," [tohu] as the proper meaning*

1 George H. Pember, *Earths Earliest Ages*, (Crane, MO: Defender Publishing, 2012), p. 21-22.

of the noun rendered "without form [bohu]." The second word signifies "emptiness," then, "that which is empty"; so that in this case the authorized translation is admissible. Now these words are found together only in two other passages, in both of which they are clearly used to express the ruin caused by an outpouring of the wrath of God.

He goes on to explain that the word "was" in this case is similar to the word "became," which means that the scripture could actually read:

"And the earth became desolate (ruin) and void; and darkness was upon the face of the deep."

Next he says,

[2]*It is thus clear that the second verse of Genesis describes the earth as a ruin; but there is no hint of the time which elapsed between creation and this ruin...*

Shiny Remnants

Did a planet in our solar system get destroyed and its matter strewn across space? Back here onEarth, we see possible fingerprints of such an event – white, shiny fragments of metal called iridium (Ir). Part of the platinum family, iridium is plentiful in meteors and comets, but is very rare on Earth – with one exception.

Most of the iridium on Earth is found in the K-T boundary, a thin layer of sediment that marks the geologic point in earth's history between the Cretaceous and Tertiary periods. Modern geology dates this period to approximately 65.5 million years ago. This is also the time period that experts say the dinosaurs went extinct because of a massive asteroid strike, specifically the strike at the Chicxulub Crater off the coast of Mexico's Yucatan Peninsula.

It is thought that this asteroid impact is what disbursed the layer of iridium across the planet. However, the amount iridium in the K-T layer is too substantial to have come from only one asteroid strike. Could it be that the K-T layer event was actually the result of multiple strikes all at the same time, or in a relative short time to one another, thereby dispersing iridium throughout the planet? If so, what was the source of all of those asteroids and comets hurling themselves at a young Earth? Is there evidence of an asteroid bombardment on Earth or elsewhere?

The moon is tidally locked to Earth so one side continually faces the blue planet while the other side is exposed to space. By studying the craters on the moon, scientists believe that the earth has been struck by at least sixty large meteors or comets three miles wide or more over the last six-hundred million years. Yet, there is no evidence of these strikes in the geological record other than the K-T boundary. Why?

---

2 Ibid.

Viewing the moon Earth's vantage point, we see some evidence of asteroid strikes but the satellite is relatively smooth. Yet when the moon is observed from the far side, it looks terribly scarred and pocked marked. If we are to assume that the earth has received that kind of punishment from celestial strikes, then perhaps barrages of large asteroids or comets have hit the earth numerous times. While it is true that earth has an atmosphere to protect it, judging by some of the larger depressions on the moon, it is reasonable to suppose that many larger strikes had penetrated the Earth's atmosphere.

Consider this: what if the massive known strikes to have it in Free State, South Africa; Ontario, Canada; Siberia, Russia; and in South Australia, just to name a few, all occurred at the same time that the Chicxulub event occurred? This heavy bombardment would not only would have killed the dinosaurs, but utterly destroyed the planet. And what if the timeframe given to such an event was far overestimated by science? What if the event occurred thousands of years ago and not millions?

Scientific Dating Methods

No doubt some readers might vehemently disagree with the last statement, but let's take a deeper look at scientific dating methods. While we do, keep in mind that a scientific theory is simply an educated guess. Theories, scientific or otherwise, are guesses until they are proven and only then they become fact. Unsubstantiated speculation - like the Oort Cloud theory - can be presented by anyone. Just because science adopts a theory and promotes it as fact, doesn't make it true. Such is the case with scientific dating methods, as the following examples illustrate:

The Geologic Table - In his excellent paper "Ten Misconceptions about the Geologic Column", Dr. Steven A. Austin points out,

> [3]*Before radiometric dating was devised, uniformitarian geologists postulated "periods" of millions of years duration to slowly deposit the strata systems. A single sedimentary lamina, or bed, was supposed by uniformitarian geologists to represent typically a year or many years duration. It was concluded, therefore, that multiplied thousands of laminae and beds superimposed required millions of years. Recently, however, geologists have discovered that laminae and beds form quickly on floodplains of rivers during floods, in shallow marine areas during storms, and in deep water by turbidity currents. The evidence of rapid sedimentation is now so easily recognized that geologists observing a strata system these days often ask where to insert the "missing time" of which the strata do not show sedimentary evidence. Catastrophism, quite naturally, is making a come-back. There is good reason to believe that entire strata systems, and even groups of systems, were accumulated in a hydraulic cataclysm matching the description of Noah's Flood in the Bible.*

3 Steven A. Austin, Ph.D., "Ten Misconceptions about the Geologic Column", Institute for Creation Research, http://www.icr.org/article/ten-misconceptions-about-geologic-column, (1984)

The point being, they are guessing at the table! No one was there. The process of setting a geologic timeline involves so many unverifiable factors that it reduces the supposed dates of millions of years of sedimentary time to simple guess work at best!

Radiometric Dating - It is assumed that this scientific method in particular is nearly foolproof. However, as Tas Walker of Creation Ministries International points out:

> [4]*It may be surprising to learn that evolutionary geologists themselves will not accept a radiometric date unless they think it is correct—i.e. it matches what they already believe on other grounds. It is one thing to calculate a date. It is another thing to understand what it means.*

Radiometric Dating is only a tool. As with all other tools, the results it gives are subject to the biases of the one who is using it.

As much as we would like to think that science has all the answers, it certainly does not. Therefore, one must use a preponderance of empirical evidence to come to educated conclusions even if those conclusions seem outlandish to the modern scientific mind.

In addition, could a Pre-Adamic earth have had another means of destruction? There are ancient texts that allude to this as well.

Before the Darkness…

This excerpt below is from the ancient writing of the Mahabharata which describes what many experts theorize could have been some kind of nuclear advance toward a prehistoric people. The Mahabharata is one of many ancient writings of this type that has been estimated as early as 400 B.C. Along with tales of ancient war and lore of fallen kingdoms, many of their current philosophical and devotional principles have derived from this work as well. The Mahabharata speaks of ancient flying machines, called Vimanas, that have the power to hurl projectiles at the earth, whose destruction lasts far beyond the initial blast.

> [5]*…a single projectile*
> *Charged with all the power of the Universe.*
> *An incandescent column of smoke and flame*
> *As bright as the thousand suns*
> *Rose in all its splendor…*
> *…it was an unknown weapon,*
> *An iron thunderbolt,*
> *A gigantic messenger of death,*
> *Which reduced to ashes*
> *The entire race of Vrishnis and the Andhakas.*

---

4 Tas Walker, "The way it really is: little-known facts about radiometric dating", Creation.com, http://creation.com/the-way-it-really-is-little-known-facts-about-radiometric-dating, (2013)

5 Philip Coppens, "Ancient Atomic Wars: Best Evidence?", *Bibliotecapleyades Online*, http://www.bibliotecapleyades.net/ancientatomicwar/esp_ancient_atomic_07.htm, (January 2005).

...The corpses were so burned
As to be recognizable.
The hair and nails fell out;
Pottery broke without apparent cause,
And the birds turned white.
After a few hours
All foodstuffs were infected...
...to escape from this fire
The soldiers threw themselves in streams
To wash themselves and their equipment.

The Knowledge Humans Were not Supposed to Receive

Among the other taboo subjects that man was not supposed to learn, historical documents show that the information given on the Vimana was actually quite mechanical and technical. These interplanetary machines were described in many ancient texts, such as Yajurveda, Mahabharata, Ramayana, Rigweda, Samarangana Sutradhara, and many others in much detail. While we have yet to recover remains of one of these crafts, they are said to be capable of traveling across air, ground, sea, and even through space.

The Samarangana Sutradhara actually goes on for 230 stanzas explaining how to construct one of these machines, listing which materials to use, and what installations are preferable for issues like avoiding fires or absorbing light and heat. It cites materials such as mercury as part of the main power source. The Samarangana Sutradhara is an operator's manual as well, explaining how to start the machine, and what to do in an emergency, such as a collision with a bird while in flight, or how to handle a malfunction and land safely.

The information available is stunning: there are directions on everything from how to disguise the ship as if in a cloud, to how to make people in enemy planes lose consciousness, for making the ship become invisible, and for tracking other aircraft in the sky. It tells which clothing pilots should wear, and how to purify the metals used in its construction. The list goes on and on.

Vimanas were able to move any direction, including diagonal, and could be used for moving people from place to place or engaging in warfare. In a similar ancient writing, the Samarangana Sutradhara, a Vimana is described as below:

> [6]*Strong and durable must the body of the Vimana be made, like a great flying bird of light material. Inside one must put the mercury engine with its iron heating apparatus underneath. By means of the power latent in the mercury which sets the driving whirlwind in motion, a man sitting inside may travel a great distance in the sky. The movements of the Vimana are such that it can vertically ascend, vertically descend, move slanting forwards or backward. With the help of the machines, human beings can fly in the air, and heavenly beings can come down to earth.*

6 "What Powered the Vimana, the 6,000-year-old Flying Machines of Ancient India?" *Ancient Code Online*, http://www.ancient-code.com/powered-vimana-6000-year-old-flying-machines-ancient-india/, (May 2016).

So not only were these vehicles presumably used to transport people, they were also bringing "heavenly beings down to earth." Since the fallen angels at this point were using interplanetary locations as a means for their dwelling place, this eludes to endless speculation on how much interaction people were regularly having with fallen angels

It is also said in the Law of the Babylonians:

> [7]*The privilege of operating a flying machine is great. The knowledge of flight is among the most ancient of our inheritances. A gift from 'those from upon high'. We received it from them as a means of saving many lives...*

Ironically, all knowledge imparted is constantly reported to be for the betterment of mankind, or as above, to save lives, but as shown in Enoch 8:12 below, the first skills listed of what was imparted seem to be preparing man for warfare.

> *...taught men to make swords, and knives, and shields, and breastplates, and made known to them the metals of the earth and the art of working them, and bracelets, and ornaments, and the use of antimony, and the beautifying of the eyelids, and all kinds of costly stones, and all colouring tinctures. And there arose much godlessness, and they committed fornication, and they were led astray, and became corrupt in all their ways...[one fallen angel] taught enchantments, and root-cuttings... [another fallen angel taught] the resolving of enchantments... astrology...the constellations...the knowledge of clouds...the signs of the earth... the signs of the sun...the course of the moon...*

As it stood, a Vimana was nothing to trifle with. It was an ancient machine of mass destruction. Its travel was charted using words that our modern science can liken to our current measurement of light years. It was capable of bringing swift destruction as shown below in the passage from the Mahabharata:

> [8]*The earth shook, scorched by the terrible heat of this weapon. Elephants burst into flames and ran to and fro in a frenzy, seeking a protection from terror. Over a vast area other animals crumpled to the ground and died. The waters boiled, and the creatures residing therein also died. From all points of the compass arrows of the flame rained continuously.*

In fact, many scientists believe that our present day deserts are actually the remnants of this warfare. Dr. J Robert Oppenheimer himself is rumored to have believed that his atomic bomb was not the first the world had seen. This matter has been the subject of many heated debates, but consider these commentaries along with the following from the Ramayana:

---

7 Ibid.

8 Monty Aldone, "Evidence for Ancient Nuclear War on Earth," *Apparently Apparel News Online*, http://www.apparentlyapparel.com/news/evidence-for-ancient-nuclear-war-on-earth, (March 28, 2014).

[9]It was a weapon so powerful that it could destroy the earth in an instant. A great soaring sound in smoke and flames And on its sits death...

Countless other ancient writings refer to cataclysmic or disastrous events, wherein the sky rains fire, and the earth suffers a colossal amount of damage. Remember the description of the destruction of Sodom and Gomorrah.

A recent article regarding the warfare described in the Mahabharata pointed out:

[10]Sanskrit scholars could not comprehend what was being described in the Epics until the dropping of the first atomic bombs on Japan...

And then later goes on to state,

[11]...the Gods were using weapons of high destructive power similar to that of nuclear power. However, it must be noted that the ancient weapons did not utilize the same earthly materials as its contents.

The Chakra Mantra

The author of an ancient text in the Mahabharata tells that knowledge was revealed to him, but says specifically that this knowledge must only be used for the good of mankind. He then discloses that unto him has been given mantras and key-words that, when used, would summon energy forms and physical weapons that could be commanded and directed: they could be propelled, stopped, or even prevented before causing destruction, only by knowing the proper mantras for circumventing the attack. These weapons could not be built using indigenous materials to earth. Their only means of creation or avoidance was by using the summoning words. On an interesting side-note: recall in Enoch above, how one fallen angel was the teacher of "enchantments," and one taught "the resolving of enchantments." It indicates that from the beginning, these forces, giving these skills to man, "for the betterment of mankind," were actually creating strife and confusion.

According to the author of this antique document, when the appropriate mantra was chanted, immense heat waves would begin to form high in the sky, creating a thick, dense layer of heat near the earth's atmosphere. These rays would continue to build until they burst, plummeting down to the earth below, creating what looked like a huge explosion at ground level. Consequently, in order to stop the attack, one must only utter the correct words for the imminently looming cloud creation to simply disperse.

There are nay-sayers today who doubt that a Vimana would have been capable of flying. But if the Chakra mantra is any indicator of the types of power they had access to just by

9 Ibid.

10 Ahmad, edited by James Hartman, "Ancient Weapons of Mass Destruction & The Mahabharata," *World mysteries Blog Online,* http://blog.world-mysteries.com/science/ancient-weapons-of-mass-destruction-and-the-mahabharata/, (June 30, 2011).

11 Ibid.

emitting mere words, then the power of levitation associated with these vehicles was most undoubtedly more than just simple mechanics.

This writing also explains that elemental events such as drought, thunder, tsunamis, hurricanes, and other natural disasters were the act of punishments by the gods. War, over-population and famine were by-products of this penalty.

Science Fiction or Non-fiction?

It may seem like all of this is science fiction until one begins to look at the indication left on earth of these long-ago events. It's easy to dismiss ancient writings such as these with a waive of one's hand in the absence of other confirmation, but once evidence is presented to back the concept of ancient nuclear war, it becomes harder to ignore.

The mysterious necklace found in King Tut's Tomb baffled Italian mineralogist Vincenzo de Michele in 1966 when it was found. He had it tested, presumably expecting it to be some sort of mineral or stone. But instead, it was glass. And not just any glass, but apparently derived from an enigmatic source. Similar glass was found intermingled with sand in the Sahara Desert, but that still didn't explain the origin of the actual glass itself. Eventually after much testing, Austrian astrochemist Christian Koeberl said that the material had been created at temperatures so high that it could only have been formed by a meteor colliding with the earth, except there was no evidence of impact. Mysteriously, when the first atomic bomb (that is to say, of this age…) was tested in 1945 in New Mexico, it left a thin layer of similar glass across the desert sand. Of this, Netscape editors said,

> [12]*But the area of glass in the Egyptian desert is vastly bigger…Whatever happened in Egypt must have been more powerful than an atomic bomb.*

And the stone in Tut's necklace is a souvenir of that explosion.

This type of glass is found throughout various deserts in the world and are known as "Seas of Glass." The only way for this glass to form is through intense heat, fusing the sand into glass. A meteor seems the most feasible explanation, until one considers the lack of impact evidence on these places of the earth.

But even more fascinating, the moon has at least one sea of glass on it as well. The moon is covered in scars and craters, indicative of explosions either at or near its surface. It also shows stress fracturing at its surface, creating what appear to be mountain ranges similar to earth's, which also show that the moon has undergone trauma, probably near the same time that earth saw hers.

In Rajasthan, India, a new living community was built over a three-square mile area. Once people started moving in, however, they started getting suspiciously ill and even developed cancer or, in the unborn, birth defects and they were left with no option but to

---

12 Netscape Editors, "King Tut's Necklace: Astonishing Mystery," *Netscape News Online*, http://channels.isp.netscape.com/whatsnew/package.jsp?name=fte/necklace/necklace&floc=wn-nx, (2016)

vacate. As the area was evacuated, and the government restricted access to the vicinity, it was discovered that there were inexplicably high radiation levels in a circular shape over the ground, with no point of impact. This pattern is characteristic of air-burst nuclear discharge.

What they found at nearby Harappa and Mohenjo-Daro really blew their minds. [13]Under the earth was evidence of what appeared to be a nuclear blast, dating back between 8,000 and 12,000 years, which apparently destroyed a prehistoric city and killing a half a million people. Skeletons were aged thousands of years because of radiation level fifty times higher than any possible natural explanation could have yielded. These were found lying in the street, many holding hands, as if they were trying to escape a very sudden doom. There was, however, no sign of actual violence regarding their deaths.

After attempting to translate the etchings within what he suspected were their temples, archaeologist Francis Taylor noted that the people probably prayed to their gods to be spared from the coming destruction. He then said,

> [14]*It's so mind-boggling to imagine that some civilization had nuclear technology before we did. The radioactive ash adds credibility to the ancient Indian records that describe atomic warfare.*

They have also found what scientists are calling "black stones;" pottery fragments bonded together, seemingly by extraordinarily intense heat.

But the evidence for such activity is world-wide. The K-T Boundary, which will be spoken about in more detail in a moment, includes a layer of ash and clay that covers the earth, on each and every continent and deep beneath the surface;

> [15]*And within this ash and clay are a preponderance of...tektites, shocked quartz, and/or glassy gobules of fused sand/dirt, the kind of which are produced by the intense heat of either a meteor strike, or a nuclear detonation...and many of them are slightly radioactive still...*

What really puzzles is the fact that this layer also contains a large amount of iridium, something that is often found in meteorites, but is extremely scarce on our planet. It can be used in nuclear reactors, as well as can be produced as a byproduct in nuclear explosions, but is so extremely rare within the earth that its availability is prohibitive. So how did earth get a substantial layer of it, world-wide, that far beneath the surface?

There are two possible answers to this question. The first is ventured by Frederick Soddy, a former Professor of Chemistry, Physicist, author and expert in radioactive elements, matter and energy. On this matter he stated:

---

13 Philip Coppens, "Ancient Atomic Wars: Best Evidence?" *Bibliotecapleyades Online*, http://www.bibliotecapleyades.net/ancientatomicwar/esp_ancient_atomic_07.htm, (January 2005)
14 Bryant Stavely, "Ancient City Found In India Irradiated By Nuclear Blast 8,000 Years Ago…" *Rense Online; excerpted from the World Island Review*, http://www.rense.com/general3/8000.htm, (January, 1992)..
15 Ibid.

[16]I believe that there have been civilizations in the past that were familiar with atomic energy, and that by misusing it they were totally destroyed.

Beyond this, the mysteries surrounding this topic go much further than the earth.

Science Myth vs. Fact

Bode's Law is a hypothesis that states that the bodies in some orbital systems, including the Sun's, orbit at semi-major axes in a function of planetary sequence. The formula suggests that, extending outward, each planet would be approximately twice as far from the Sun as the previous planet. The equation applies to our own solar system nearly perfectly and helped locate Uranus and the planetoid Ceres. However, there is one place in this planetary system where the equation should work, but doesn't. Maybe...

According to Bode, there should be a planet in between Mars and Jupiter. Instead, what we find is an asteroid debris field dubbed the Asteroid Belt which runs approximately 500 miles across, its girth clustered with planetismal-sized chucks, its culminations diminishing down to fine dust. Scientists say that these loose pieces of space rock are the precursors to a planet that was never fully formed in the early development of the solar system. These big brains say that the gravitational pull from Jupiter so greatly hindered this wouldbe planet, that rather than coming together in celestial evolution, the pieces just hung back and scattered in this tight area.

However, if you were to gather all the asteroids together from the belt and glue them into a ball, it would be smaller than the size of our own moon - Not very planetary. And if you consider that the Asteroid Belt is one-half the distance from the sun to Jupiter and the sun has a 1000 times stronger gravitational pull than Jupiter, which has significant gravity, since Jupiter is closer to Saturn, shouldn't Saturn be the planet that didn't develop? Yet, like so many unproven scientific theories that become like religious dogma, this idea about the lack of a developed planet in this area persists.

Proof of a destroyed planet is everywhere throughout our solar system, from the variable degrees that planets' axis angle at, to scarring on the surfaces of celestial bodies, and even the rings on planets. It is manifest in the fact that astral bodies showing signs of ancient vegetation and civilization are now deserted and incapable of habitation. Further evidence in our solar system can be found with comets, as CK Quarterman illustrates below:

[17]Comets provide the most convincing argument in favor of a destroyed planet. Comets cannot date back to the beginning of the solar system because they consist of evolved matter such as water which cannot form in the void of space. Water can only form on the surface of a planet with an atmosphere.

---

16 Ibid.

17 CK Quarterman, "Scientists Find Evidence of Home of Fallen Angels," http://www.fallenangels-ckquarterman.com/scientists-find-evidence-of-rahab-lucifers-home-planet/, (November 11, 2011).

The aptly named Rahab, which meant "boaster, or pride" was the home of Satan and the fallen angels and known to be the "stone of fire." Remember that the Vimanas were said to bring the heavenly beings down to earth. These interplanetary vehicles surely played a large role in the fallen angels' interaction with mankind, and since Rahab was a planet probably able to sustain life, people may have been taken up to the planet as well. Since the ruins of the planet show that it had a high content of diamond and ore, it's possible humans were even enslaved and forced to labor on it.

We live in a solar system that should have ten planets. Rahab was planet number five, which should be between Mars and Jupiter. The asteroid belt mentioned before is all that is left of this destroyed world.

If you think these statements are purely myths developed in this author's mind, let us take a look at another "scientific" myth that, with much less evidence, is considered fact.

The Oort Cloud is a hypothetical extended shell of icy objects that is speculated to exist in the outermost reaches of the solar system. It's named after a Dutch astronomer named Jan Oort, who first theorized its existence in 1950. According to the scientist the Oort Cloud is roughly spherical and most of the long-period comets in our solar system have originated from this area.

But here's the problem – Even with the Hubble Telescope, the planetoid Pluto is only a fuzzy blob. The Cloud supposedly exists at 50,000 Astronomical Units (AU) away and has never been seen and not one shred of physical evidence has been obtained as to its existence. Yet, this myth is declared as fact in mainstream astronomy.

Comets in our solar system originate here? Let's test this idea. Comets were thought to be dirty balls of ice that, as they got closer to the sun, began to melt. This melting is what caused the tail of the comets to brighten while in close proximity to sun according astronomers.

However, on July 4th 2005 an impact probe from the Deep Impact mission was launched from the spacecraft in order to hurl itself into comet Tempel 1. What was found is surprising. Although the comet did have ice on it, it was primarily made up of silicates, clay, carbonates, iron-bearing compounds, aromatic hydrocarbons and high concentrations of iridium. Keep this last item in mind as we go forward.

When the probe smashed into the comet, it generated a huge flare that caught the scientists off guard. It is surmised that this was an electrical reaction to the compounds, particularly the iridium, which caused the flare. Therefore, one has to ask, could these bodies be electrical, or electric-magnetic, pieces of space rock and did they inherit these properties from somewhere else? This exercise has also taught science to think differently about what causes the tail of a comet to light up like as it goes around the sun.

As mentioned earlier, meteors also have high concentrations of iridium and appear to have electro-magnetic properties. Could it be that both comets and meteors within our solar system originated from the same place? Could it be that that place is not from some fictitious Oort Cloud on the edge of the solar system, but is someplace much closer to home? Could it be that the inertia that propelled these comets into high-angled hyperbolic orbits was from an extreme collision that decimated a planet that, according to Bode's Law, should be between Mars and Jupiter? Could it be that the asteroids and dust that remain in the Asteroid Belt are the leftovers of this violent collision that pulverized a shiny iridium based planet? Is there physical evidence in the solar system to suggest that such a collision took place? Finally, was that planet inhabited and, for our purposes, was it destroyed for a reason? Like – rebellious angels?

Science Myth vs. Fact

Bode's Law is a hypothesis that states that the bodies in some orbital systems, including the Sun's, orbit at semi-major axes in a function of planetary sequence. The formula suggests that, extending outward, each planet would be approximately twice as far from the Sun as the previous planet. The equation applies to our own solar system nearly perfectly and helped locate Uranus and the planetoid Ceres. However, there is one place in this planetary system where the equation should work, but doesn't. Maybe…

According to Bode, there should be a planet in between Mars and Jupiter. Instead, what we find is an asteroid debris field dubbed the Asteroid Belt which runs approximately 500 miles across, its girth clustered with planetismal-sized chucks, its culminations diminishing down to fine dust. Scientists say that these loose pieces of space rock are the precursors to a planet that was never fully formed in the early development of the solar system. These big brains say that the gravitational pull from Jupiter so greatly hindered this wouldbe planet, that rather than coming together in celestial evolution, the pieces just hung back and scattered in this tight area.

However, if you were to gather all the asteroids together from the belt and glue them into a ball, it would be smaller than the size of our own moon - Not very planetary. And if you consider that the Asteroid Belt is one-half the distance from the sun to Jupiter and the sun has a 1000 times stronger gravitational pull than Jupiter, which has significant gravity, since Jupiter is closer to Saturn, shouldn't Saturn be the planet that didn't develop? Yet, like so many unproven scientific theories that become like religious dogma, this idea about the lack of a developed planet in this area persists.

Proof of a destroyed planet is everywhere throughout our solar system, from the variable degrees that planets' axis angle at, to scarring on the surfaces of celestial bodies, and even the rings on planets. It is manifest in the fact that astral bodies showing signs of ancient vegetation and civilization are now deserted and incapable of habitation. Further evidence in our solar system can be found with comets, as CK Quarterman illustrates below:

[18]Comets provide the most convincing argument in favor of a destroyed planet. Comets cannot date back to the beginning of the solar system because they consist of evolved matter such as water which cannot form in the void of space. Water can only form on the surface of a planet with an atmosphere.

The aptly named Rahab, which meant "boaster, or pride" was the home of Satan and the fallen angels and known to be the "stone of fire." Remember that the Vimanas were said to bring the heavenly beings down to earth. These interplanetary vehicles surely played a large role in the fallen angels' interaction with mankind, and since Rahab was a planet probably able to sustain life, people may have been taken up to the planet as well. Since the ruins of the planet show that it had a high content of diamond and ore, it's possible humans were even enslaved and forced to labor on it.

We live in a solar system that should have ten planets. Rahab was planet number five, which should be between Mars and Jupiter. The asteroid belt mentioned before is all that is left of this destroyed world.

If you think these statements are purely myths developed in this author's mind, let us take a look at another "scientific" myth that, with much less evidence, is considered fact.

The Oort Cloud is a hypothetical extended shell of icy objects that is speculated to exist in the outermost reaches of the solar system. It's named after a Dutch astronomer named Jan Oort, who first theorized its existence in 1950. According to the scientist the Oort Cloud is roughly spherical and most of the long-period comets in our solar system have originated from this area.

But here's the problem – Even with the Hubble Telescope, the planetoid Pluto is only a fuzzy blob. The Cloud supposedly exists at 50,000 Astronomical Units (AU) away and has never been seen and not one shred of physical evidence has been obtained as to its existence. Yet, this myth is declared as fact in mainstream astronomy.

Comets in our solar system originate here? Let's test this idea. Comets were thought to be dirty balls of ice that, as they got closer to the sun, began to melt. This melting is what caused the tail of the comets to brighten while in close proximity to sun according astronomers.

However, on July 4th 2005 an impact probe from the Deep Impact mission was launched from the spacecraft in order to hurl itself into comet Tempel 1. What was found is surprising. Although the comet did have ice on it, it was primarily made up of silicates, clay, carbonates, iron-bearing compounds, aromatic hydrocarbons and high concentrations of iridium. Keep this last item in mind as we go forward.

When the probe smashed into the comet, it generated a huge flare that caught the

18 CK Quarterman, "Scientists Find Evidence of Home of Fallen Angels," http://www.fallenangels-ckquarterman.com/scientists-find-evidence-of-rahab-lucifers-home-planet/, (November 11, 2011).

scientists off guard. It is surmised that this was an electrical reaction to the compounds, particularly the iridium, which caused the flare. Therefore, one has to ask, could these bodies be electrical, or electric-magnetic, pieces of space rock and did they inherit these properties from somewhere else? This exercise has also taught science to think differently about what causes the tail of a comet to light up like as it goes around the sun.

As mentioned earlier, meteors also have high concentrations of iridium and appear to have electro-magnetic properties. Could it be that both comets and meteors within our solar system originated from the same place? Could it be that that place is not from some fictitious Oort Cloud on the edge of the solar system, but is someplace much closer to home? Could it be that the inertia that propelled these comets into high-angled hyperbolic orbits was from an extreme collision that decimated a planet that, according to Bode's Law, should be between Mars and Jupiter? Could it be that the asteroids and dust that remain in the Asteroid Belt are the leftovers of this violent collision that pulverized a shiny iridium based planet? Is there physical evidence in the solar system to suggest that such a collision took place? Finally, was that planet inhabited and, for our purposes, was it destroyed for a reason? Like – rebellious angels?

Rahab

The second answer to how the earth got a substantial world-wide layer of iridium beneath surface might also be found in the bible itself. However, first we must look at what the Heavy Bombardment Period was and what were the results. According to the BBC:

> [19] *The Late Heavy Bombardment (abbreviated LHB and also known as the lunar cataclysm) is an event thought to have occurred approximately 4.1 to 3.8 billion years (Ga) ago, at a time corresponding to the Neohadean and Eoarchean eras on Earth. During this interval, a disproportionately large number of asteroids are theorized to have collided with the early terrestrial planets in the inner Solar System, including Mercury, Venus, Earth, and Mars. The LHB happened after the Earth and other rocky planets had formed and accreted most of their mass, but still quite early in Earth's history.*
>
> *Evidence for the LHB derives from lunar samples brought back by the Apollo astronauts. Isotopic dating of Moon rocks implies that most impact melts occurred in a rather narrow interval of time. Several hypotheses attempt to explain the apparent spike in the flux of impactors (i.e. asteroids and comets) in the inner Solar System, but no consensus yet exists.*

There is ample evidence to suggest that it was this cosmic rock storm that deposited iridium throughout the planet. Therefore, for the sake of discussion, let's assume that it was these multiple asteroid impacts, all at the same time, that deposited the iridium in the K-T layer. If we do, then it appears likely that the source of those impacts can be tied

19 "The Late Heavy Bombardment Ends," BBC, http://www.bbc.co.uk/science/earth/earth_timeline/late_heavy_bombardment, (2016)

directly to something called Rahab as read in Job 26:11-12:

The pillars of heaven tremble and are astonished at his reproof. He divideth the sea with his power, and by his understanding he smiteth through the proud.

This is the Kings James Version, and it does no justice to the original language. In Hebrew, verse 12 literally says: (*raga*) (*hai yam*) (*be cho chov*) (*u vit vu na tov*) Rahab. *Which means* (to disturb) (the sea) (with His power) (and understanding or discernment) (**to shatter**) a (boasting prideful) Rahab.

This is very interesting because we see through Earth's own history that God brought destruction to this planet with – the sea in the Flood. As Solomon said, "nothing new under the sun." Yet, with the Earth, the Creator showed restraint even in bringing the Flood. The occupants of Rahab sins were so egregious; God absolutely "shattered" it as Psalm 89:10 says:

Thou hast broken Rahab in pieces, as one that is slain; thou hast scattered thine enemies with thy strong arm.

We also see that it was full of His enemies. Rahab is not simple pride as translated in the King James translation of Job 26:11-12, but appears to have been a place from where God scattered His enemies.

Rahab comes into focus with Isaiah 51:9:

Awake, awake, put on strength, O arm of the LORD; awake, as in the ancient days, in the generations of old. Art thou not it that hath cut Rahab, and wounded the dragon?

There are several important points in this verse:

- First – The book of Isaiah was written somewhere between 760 and 700 BC,

- Second – The verse is referring to "as in ancient days" and "generations of old". The word "ancient days" is from the Hebrew *qedem* (קֶדֶם), Strongs #6924a, and literally means earliest times. "Generations of old" is our word olam again which can mean antiquity, but can also mean ages, or eternity. Consider that the oldest patriarch of Israel, Abraham, lived during the Isin Babylonian Dynasty which was in power from 2017 to 1787 BC. Are we to believe that Isaiah is referring to a thousand years in the past as being "ancient days" or ages, or is he speaking of a time even before Abraham? Adam would have been created nearly 3000 years before Isaiah was born, so the prophet could have been referring to this timeframe. However, there is no account in scripture during either Abraham or Adam's time of a Rahab being "cut". Therefore, one can only assume that Isaiah is referring to another time. Since Adam was the beginning of mankind, it must have been a time before that.

- Third – The verse says that God "wounded the dragon". There is another important descriptor for "the dragon" in Revelation 12:9. It is none other than Satan himself.

It appears that this verse is speaking about an ancient time even before Adam and Eve – a Pre-Adamic era when God wounded Satan and scattered those that were with him. If so, where was this place called Rahab that was cut, or shattered or, as one version translates Job 26:11, "pulverized" by God's rebuke? It could be referring to Earth during the period when the K-T layer was formed, as asteroids rained down on the planet and reduced it to what was probably a pre-creation state. However, earth wasn't shattered or cut to pieces and this terminology denotes total destruction.

As mentioned in Chapter 1, God said in Jeremiah 4:27 that He would not make earth's punishment "a full end" meaning that He would not totally destroy it. The references to Rahab indicate a "full end", that it was totally destroyed. Therefore, it is logical to deduce that Rahab is not earth, but somewhere else.

But this judgment of the dragon would reach earth because God purposed it as a witness to others. This is the message of Ezekiel 28:18:

> *Thou hast defiled thy sanctuaries by the multitude of thine iniquities, by the iniquity of thy traffick; therefore will I bring forth a fire from the midst of thee, it shall devour thee, and* ***I will bring thee to ashes upon the earth in the sight of all them that behold thee.*** *[Emphasis Added]*

What were the ashes that rained down on the earth for all to see? Perhaps they were the fragments of a planet that no longer exists.

[20]The Southwest Research Institute (SRI) in San Antonio, Texas has done statistical analysis on the likelihood of our solar system only having four large planetary giants, Jupiter, Saturn, Uranus, and Neptune. Their models indicate that a fifth planet was most likely part of the solar system at one time. Could this planet be the source of the iridium in the Earth's K-T layer? If so, then we can surmise that the place God destroyed because of the rebellious dragon and his followers was none other than the original fifth planet from the sun – Rahab.

[21]Just as an expert chess player sacrifices a piece to protect the queen, the solar system may have given up a giant planet and spared the Earth…

But what was this other, mystifyingly lost planet?

---

20 CK Quarterman, "Rahab the Home of Fallen Angels," http://www.fallenangels-ckquarterman.com/rahab/, (October 20, 2011)

21 "Giant planet ejected from the solar system," *Southwest Research Institute News Online*, http://www.swri.org/9what/releases/2011/giant-planet.htm, (November 10, 2011).

Rahab would have been enormous and big enough to hold ten Earths. It would have also dominated Earth's night sky and probably was visible during the day as well. The type of fragmentation and liquid content in its ruin are indicators that it was a solid, water-bearing planet, and probably habitable.

As previously stated, Rahab meant "boaster, or pride" and was the home of Satan and the fallen angels and known to be the "stone of fire." Remember that the Vimanas were said to bring the heavenly beings down to earth. These interplanetary vehicles surely played a large role in the fallen angels' interaction with mankind, and since Rahab was a planet probably able to sustain life, people may have been taken up to the planet as well. Since the ruins of the planet show that it had a high content of diamond and ore, it's possible humans were even enslaved and forced to labor on it. In addition, prior to Rahab's destruction the angels who had left their first estate, clipped-winged-angels, also used stargates to get off planet and make their home in other parts of the solar system.

This planet's destruction was so thorough that the angelic civilization on Mars was also destroyed. The fallout spread across the solar system, including Earth, which was bombarded with asteroids in an Extinction Level Event. This barrage not only killed the dinosaurs but also erased the angelic Pre-Adamic civilization that existed on this planet.

Regarding those asteroids and comets, it is highly likely that these space rocks maintained part of their unique electromagnetic charge that they once held as part of the whole planet of Rahab. When we see a flare as a result of the Deep Impact Mission mentioned earlier, it should not be surprising that the static discharge occurred. Mike Wall of Space.com wrote an article about how asteroids retain a magnetic field.

> [22]*The magnetic fields of these big asteroids were apparently generated by the same process that drives Earth's global magnetic activity, and could have persisted for hundreds of millions of years after the objects' formation, researchers said.*

Such destruction wrought on Rahab and the solar system should give every twenty-first century human pause and cause them to ponder if this kind of terror could be revisited upon this solar system and Earth in the future. The answer to that question would be a sorrowful yes. It happens at least once every 3,600 years, according to the Sumerians. And we are due.

Scarred Remains

In the last chapter we noted how the planet Mars had been knocked off its axis. To this point, David Flynn explains:

> [23]*The face on Cydonia, was once on the Martian equator aligned upright to the north, but presently is located at 41° north latitude, tilted from upright (north-*

22 Mike Wall, "Magnetic Fields of Asteroids Lasted Hundreds of Millions of Years," Space.com, http://www.space.com/28319-asteroid-magnetic-fields-earth-core.html, (January 21, 2015)
23 Ibid.

oriented) by an angle of 22°. Some cataclysmic event has occurred since the building of the Cydonia monuments on Mars, which destroyed Martian oceans and the atmosphere, and also tilted its axis of rotation.

In addition, there are numerous pockmarks on the surface that indicate a pelting of space debris. Mars also has a tidally locked moon, Phobos, which, like our moon is smooth on the side facing the planet. However, the space side of Phobos is even more scarred than our own moon and appears to be burnt black, indicating a superheated assault on the satellite. Taken together, these features lead to speculation that a massive series of impacts from space may have suddenly and violently ripped away the atmosphere of Mars.

Because of the constant cloud cover of both Jupiter and Saturn, evidence of impacts on those spheres are hidden from view. However, the sum of the evidence presented here provides enough reason to propose that there was at one time a planet residing in the location between Mars and Jupiter, and that this planet was suddenly and violently pulverized.

If this how this planet met its fate, could such violence in our solar system or galaxy happen again? Was this a form of God's judgment? And if so, why?

Whatever Was, Will Be

If you are beginning to sense an uncomfortable pattern of repetition, it is understandable. The behavior of those angels that left their first estate hasn't changed in thousands of years. They corrupted all flesh, both spiritually and genetically in Genesis 6. Going further back, Satan in the Garden twisted God's words and talked Adam and Eve into sinning. Yes, they did have a choice, but the Enemy showed just how shrewd he was.

We can surmise from the Heavy Bombardment Period that his actions and his fellow fallen angels were up to their old tricks which were so egregious that the Creator decided to wipe out their habitats on Rahab and Mars, and most likely on Earth. At least that is the picture that is coming into focus as we put this puzzle together.

Yet, you could be sitting back thinking this is pure conjecture; that we have science and history to gain understanding from. Why reduce ourselves to speculation and fantasy? If you fall into this camp, you would be hazardously wrong. Why?

Because if you understand anything about God, it is the fact that He is consistent. If through Solomon He told us that we would see repeating patterns on this earth – then we are seeing repeating patterns. The difference of course is that we live in the 21st century. We are so intelligent and we have technology after all.

Therein lies the problem – the fallen angels had technology and they had the same kind of hubris.

Whatever was, is what will be.

Biographies

Donald W. Patten

Donald Wesley Patten was born in 1929 at Conrad, Montana. He received a BA from the Geography Department, Univ. of Washington in 1951, and also the MA degree in Geography in 1961. He is married, has seven children and sixteen grandchildren. In Seattle, he was the founder of Microfilm Service Co., and also of Patten Corporation, a firm involved in commercial property management.

He has written three books, "The Biblical Flood and the Ice Epoch" (1966), "The Long Day of Joshua and Six Other Catastrophes" (1973) and "Catastrophism and the Old Testament" (1988). He edited the series for Baker Book House, 1968-1980, entitled "Symposium on Creation".

With Windsor, he has co-authored "The Recent Organization of the Solar System", and nine published essays.

Patten and Windsor were among a party of ten, led by David Fasold, who inspected the Ark Site, the Chip Stone Site and the Drogue Stones Site on June 21, 1990, all near Dogubayazit, Turkey. The Ark Site is two miles from the Iranian border, at an elevation of 6,300 feet. It is 18 miles due south of Mt. Ararat.

Patten also is a principal in a current experiment on the relationship of effect of elevated carbon dioxide in controlled atmospheres on the size and longevity of mammals.

Patten is a member of Bethany Community Church in Seattle and is on the endowment committee of Shoreline Christian School (formerly Watson Groen Christian School), Shoreline, Washington. He is a member of Seattle's Ravenna Avenue Chess Club.

Samuel R. Windsor

Samuel Ronald Windsor was born in 1934 at Wichita, Kansas. He received a BS in Mechanical Engineering in 1959 at Wichita State University. He resides on Orcas Island, Washington. He is married, has nine children and sixteen grandchildren. He is the associate partner in Bronson & Windsor, a firm in naval architecture and marine engineering in Seattle's Ballard district. Windsor also is an engineer specializing in forensics, involving industrial accidents. Windsor has appeared in courts as a scientific witness in countless law suits involving industrial accidents, sometimes for the plaintiff, sometimes for the defendant. They have been high figure law suits where an engineering analysis of the accident scenes have been needed.

Windsor co-authored with Patten "The Recent Organization of the Solar System". He has also authored several essays on Earth history. These include "The Venus vs. Mars Controversy (or Which Planet Assaulted the Earth"), "Noah's Ark, 24,000 Deadweight Tons", and "Noah's Ark: Its Geometry". These were published in the periodical "Catastrophism and Ancient History". Also, he has co-authored with Patten seven other published essays on Earth history.

Windsor and his wife (Erma) were among the party of ten who, in June 1990, led by David Fasold, inspected the Ark Site, the Chip Stone Site and the Drogue Stones Site.

Windsor is a member of the International Conference of Building Officials, the American Concrete Institute and American MENSA.

CHAPTER 1

The Little Bang Theory

> *Theories of their [asteroid] origin are divided between their being (a) the debris of an ancient planetary collision, and of (b) the material of which a planet might have been made, but was not, because of the gravitational influence of Jupiter, with the preponderance of opinion favoring the latter theory.*
>
> James S. Pickering (1958)

Introduction

In the year 2003, 44 years have passed by since the assessment of asteroid origin was made by James S. Pickering, a statesman among astronomers. Since then, there have been over 50 lunar missions, over 50 solar missions and some 50 planetary missions to various planets.

Among the planet surfaces photographed were the back side of the Moon, then of Mars, Venus, Mercury and on to Jupiter, Saturn, Uranus and Neptune. Astonishing physical geography has been discovered on many of the planets and satellites, evidence in many cases of sudden, cataclysmic violence.

Has anything changed concerning theory of asteroid origin in the last 44 years? Opinion still is divided between those who feel the asteroids accreted in space, like condensing dew drops, over 4 billion years, and those who still feel that there might have been a planetary collision somewhere between Jupiter and Mars.

The consensus among astronomers is about the same today as it was 44 years ago, with the majority still being "accreters" and the minority being "collisionists with an unknown planet". The thrust of the next six chapters is to provide a final explanation for the genesis of the asteroids. A baker's dozen of categories of evidence are cited.

Twelve or thirteen items of physical evidence are presented and discussed in the next six chapters. The evidence will be from the scars of Mars and its two tiny satellites. It will be that Mars was out there, in an orbit much different than its modern orbit, and it was the red planet that caused the fragmentation of the asteroids. But that is just part of the damage caused by Mars in its former orbit.

For the demise of a small former planet, the next six chapters will convict Mars, but a Mars not in its modern orbit. It will convict a Mars having a different orbit, one long and narrow rather than an orbit almost circular. It will be an orbit where the energy of Mars in its former orbit and the energy of Mars in its recent orbit are similar. But the regions Mars roamed in space were vastly different than it is today.

This evidence, accurately reported and interpreted, will show that the "planetary collisionists", the minority of astronomers, have been closer to the truth, and the majority, the "accreters" have been dead wrong.

However, today Mars does not go out into the asteroid orbit. Most of the planetary collisionists, like the accreters, also suppose the orbit of Mars has been unchanged in the last 4 billion years. With the dogma of gradualism they too have embraced, they too have failed to suspect the obvious candidate, and the one planet closest to the asteroid belt. Why the failure? Their thinking has been from a defective paradigm.

The paradigm offered here is that the Mars did not assume its modern orbit until the year 701 B.C.E., well within the boundary of ancient writings and ancient history. If so, what was the orbit of Mars in its previous age? Chapters 7 to 12 will address this, the deeper question.

Twenty Questions

Twenty questions are broached to encourage thought. The phrase in Latin is "sapere aude", - dare to think. The dogmas of gradualism, taught over the past 150 years in the western science, frankly are wearing thin. Perhaps the dogmas of gradualism in Earth history are due to be supplanted. If so, that supplanting will be by planetary catastrophism.

ISSUE # 1. The Epic of Gilgamesh is an ancient Sumerian account of history that extends back to the third millennium B.C.E. It was written on clay tablets with wedge-shaped characters. Many of those clay tablets have been found, and have been translated into English. In over a dozen of them, there is a discussion about Noah and the Ark, and ancient survival amid holocausts of water.

In the Epic of Gilgamesh, the most complete of the cuneiform flood accounts, much veneration was given by the ancients to the planets. They included Innanna, or Venus. Ea was the Earth. Anu was Jupiter. Ninnurta was Saturn. Enlil was Mars.

Tablet 11 in the Epic of Gilgamesh gives a Sumerian account of Noah's Flood in 300 lines. It is an account four times as long as is the Genesis account of Noah's flood. It indicts Enlil-Mars as the cause of Noah's Flood.

(But) Enlil shall not come near to the offering,
Because without reflection he brought on the deluge
And consigned my people to destruction!
As soon as Enlil arrived and saw the ship,
Enlil was wroth; ... [n1]

QUESTION # 1. Why was Enlil-Mars blamed by Gilgamesh, some 200 years after the event, for the deluge?

ISSUE # 2. Ginzberg's Legends of the Jews is a Talmudic Commentary in seven volumes. The Jews absorbed considerable of the cosmology of the ancient Near East when in the Babylonian Captivity. It was from Persian sources, Mede sources, Sumerian sources, Chaldean sources, etc., sources both Semitic and Aryan. The Talmudic compendium states, quite definitely (if not thoroughly), the following:

> The flood was produced by a union of the male waters, which are above the firmament, and the female waters issuing from the earth. The upper waters rushed through the space left when God removed two stars out of the constellation Pleiades. [n2]
>
> There were other changes among the celestial spheres during the year of the flood. [n3]

QUESTION # 2. Which two stars? And what other changes? Was it changes in climate? Latitude? Topography? Orbit? Population? Place of Residence?

ISSUE # 3. In the Old Testament, catastrophes are not only a repeating scenario; they are periodic. Usually they are reported to involve earthquakes, cosmic lightning, thunderous cosmic noise, volcanism and celestial flyby scenery. Was the cause of these catastrophes astronomical or earthly?

Four times in Hebrew records, the night of March 20-21 involved an angel from the heavens bringing fire, celestial noise, earthquakes, lightning and sudden destruction. It was their "Passover."

QUESTION # 3. Could these days of upheaval have had anything to do with Mars? Was Mars the destroying angel that appeared so suddenly, and also departed so suddenly?

ISSUE # 4. In ancient Talmudic literature, there were seven archangels, or super angels. Each archangel was assigned as comptroller of one luminary, five for planets, and one each for the Sun and the Moon. One archangel, Sammael by name, was the comptroller of Mars. It was an evil angel. Was Mars as the planet like Sammael, the evil, destructive angel?

On the other hand, an old tradition states (Al-Barceloni, 247) that each one of the seven planets has its own angel as follows; the sun has Raphael; Venus, Aniel; Mercury, Michael; the moon, Gabriel; Saturn, Kafziel; Jupiter, Zadkiel; Mars Sammael. [n4]

Cain's descent from Satan, who is the angel Sammael, was revealed in his seraphic appearance. [n5]

PRE gives the downfall of Sammael and his host as the first penalty of the serpent, in agreement with the view of this Midrash, according to which the real seducer was Satan (= Sammael), who made use of the serpent ... [n6]

QUESTION # 4. What kind of evil did Sammael wreak on the Earth? See Psalm 18 or II Samuel 22.

Then the earth shook and trembled; the foundations of heaven moved and shook, because he was wroth.
There went up a smoke out of his nostrils, and fire out of his mouth devoured: coals were kindled by it.
He bowed the heavens also, and came down; and darkness was under his feet.
He rode upon a cherub, and did fly: and he was seen upon the wings of the wind.
And he made darkness pavilions round about him, dark waters, and thick clouds of the skies.
Through the brightness before him were coals of fire kindled.
The Lord thundered from heaven, and the most High uttered his voice.
And he sent out arrows, and scattered them; lightning, and discomfited them.
And the channels of the sea appeared, the foundations of the world were discovered, at the rebuking of the Lord, at the blast of the breath of his nostrils.
II Sam. 22:8 and Psalm 18:7-15 identical passages.

ISSUE # 5. Homer, a Greek writer in the 9th century B.C.E., described a fiery holocaust which enveloped the Greek encampment as it was preparing to assault Troy. Homer described Ares as "bane of mortals." In old English, "bane" meant murderer, bringer of ruin and woe, that which destroys life.

QUESTION # 5. Why did Homer ascribe such a ruinous, murderous role to the planet Ares? Did Homer really mean the planet the Romans called Mars?

ISSUE # 6. Hesiod wrote, in his "Shield of Herakles, at the end of the 8th century B.C.E., about an Ares-Athene conflict and also about an Ares-Hera war in the cosmos." While it is only 480 lines long, no less than seventeen times Hesiod refers to Deimos and Phobos in one fashion or another as the two steeds of Ares, pulling the chariot of Ares across the cosmos.

Deimos and Phobos are so small that they weren't discovered until 1877, when finally telescope technology was up to the task.

QUESTION # 6. How could Hesiod possibly have known that Mars has two tiny satellites unless he saw them circling Mars amid flyby chaos. He described both the two black steeds circling Mars and the chaos of the flyby scene.

ISSUE # 7. The ancient Romans venerated the planets, and above all Mars. For Mars adoration, they set aside two days. One was their tubulustrium, a day of trouble, turmoil and tumult, on March 20-21. The other was their armilustrium, a day of alarm, October 24.

QUESTIONS # 7. Why did their tubulustrium correlate precisely with the date of the Hebrew Passover? Why did their armilustrium correlate with October 24, the day of the onset of Noah's Flood? October 24 was the 17th day of the second month in the old Tishri calendar of the Near East. Was the Earth's orbit crossed twice by Mars, once when coming in and once again while going out? What was going on?

What was so special, and so fearful to the early Romans about Mars?

ISSUE # 8. The Earth's rotation rate is once in 1,436 minutes, as measured by the fixed stars. The spin rate of Mars is once in 1,477 minutes. The differential of the two spin rates is only 2.8%. The similarity is 97.2%. Gradualistic astronomers hold that it is a coincidence.

QUESTION # 8. Why are the spin rates of Mars and the Earth so similar? Is this a coincidence? Really?

ISSUE # 9. The spin axis tilts of Mars and the Earth also are very similar. The tilt of Mars is 23° 59 minutes while the tilt of the Earth is 23° 27 minutes. They are within 32 minutes of each other. They, too, are nearly identical, within 2.3%.

QUESTION # 9. Why are the spin axes inclinations of Mars and the Earth so similar? Is this also a coincidence? Really? If so, then it is coincidence "squared", a particularly unique degree of coincidence.

ISSUE # 10. Space probes Mariner 6, 7 and 9 flew within 2,000 miles of the surface of Mars. That was in 1969, 1969 and 1971. They photographed tens of thousands of craters on the surface of Mars. Turn one of the hemispheres of Mars so that its center is at latitude 45° S. and at longitude 320° W. Turn the other hemisphere's center to latitude 45° N. and 140° W. longitude.

From those two centers, in one hemisphere there is a count of craters, all 20 miles in diameter of more, of 3,068 (93%) and the other hemisphere has a crater count of only 237 (7%). One hemisphere is among the most badly battered, bombarded, clobbered hemispheres in the Solar System. And next to it the other hemisphere is the most serene. They are side by side on the same planet, and that planet is our near neighbor.

The Moon has been photographed now on both sides, and both sides are heavily poxed with craters, craters which have a random distribution. The craters also are randomly distributed on the satellites of Jupiter. And also on the one side of Mercury photographed to date. But not on Mars.

QUESTION # 10. Why is there such a radically different distribution of craters on the surface of Mars. And also, why would anyone in his right mind want to count the craters on Mars anyway?

ISSUE # 11. Mars has just a few volcanoes, but they are giants. The largest, Olympus Mons is 13 miles high and its volcanic cone contains 450,000 cubic miles of lava flow, ash and other ejecta. The base of this cone covers 100,000 sq. miles, like the size of the state of Arizona or of Colorado.

QUESTION # 11. Why does such a tiny planet, Mars have such gigantic volcanoes? Like Mae West, was it squeezed hundreds of times?

ISSUE # 12. Underneath the thick cloud cover of Venus is a surface that is badly battered with a variety of kinds of violence producing violent forms of physical geography. The torturing of the crust of Venus looks as if it happened just yesterday.

QUESTION # 12. Why is the crust of Venus so badly scarred? Why do the scars of Venus appear to be so fresh?

ISSUE # 13. Half of the surface of Mercury has been photographed, and the rest is in need of space probe attention. On that half, there is a flat lake made of lava more than twice the area of Texas, and several other large lava flows. But there are no volcanic cones on Mercury's known surface, and moreover, there is evidence Mercury's center is a solid, through and through, and is not fluid.

The Sun warms the surface of Mercury to a toasty 700 or 800° F. during its 88-day "day." But it requires 2,900 to 3,000° to melt crustal compounds such as silica, hematite, magnetite and alumina. In addition, Mercury has craters with walls that are partly, but not fully melted down.

QUESTION # 13. Where did the sustained heat come from, more than 3,000° F., needed to liquefy large amounts of the crust of Mercury? And for how long was that torching sustained? Did that torching affect the Earth?

ISSUE # 14. The Sun's diameter has been measured and recorded for the last 150 years at the Greenwich Observatory, London. Both the latitudinal and the longitudinal diameters have been measured annually and recorded. For this time span, the Sun has been shrinking, apparently because it is cooling. The rate of shrinkage is presently 120 feet per day. This is a shrinkage of one mile in 44 days. The Sun's diameter is about 864,000 miles.

864,000 X 44 = 38,016,000 days. This divided by 365.256 = 108,040 years.

QUESTION # 14. What does the shrinking of the Sun indicate in times past? In times future? 100,000 years ago was the Sun twice its present diameter? 100,000 years in the future will the Sun cool and shrink to become more like a planet? Or is the diameter of the Sun in some kind of an oscillating stage, and is just in a down sweep of an oscillation?

ISSUE # 15. One hemisphere of Mars has been badly battered by a sudden blizzard of fragments. It has the craters to prove it, and the other side lacks the craters also to prove it.

QUESTION # 15. Does this mean that Mars was the felon, cosmic culprit that caused an even smaller planet, somewhat farther out, to fragment into asteroids?

ISSUE # 16. The known asteroid fragments are over 5,000 in number with more being discovered each year. The largest three asteroids have diameters of 625 miles, 380 miles and 330 miles. They are Ceres, Pallas and Vesta.

QUESTION # 16. Are they remnant fragments of a small, former tenth planet?

ISSUE # 17. The closest distance to the Sun is the "perihelion" of the asteroids. It averages somewhere between 225,000,000 to 235,000,000 miles from the Sun. If the asteroids suddenly became fragments, it would have been the biggest "little bang" in the history of the Solar System. Whatever occurred, it was roughly 225,000,000 miles from the Sun.

QUESTION # 17. Is this evidence? Was Mars formerly once out there in a different orbit?

ISSUE # 18. All students of science have been taught that the orbits of Mars and the other eight planets were set in cement 3 or 4 billion years ago. No orbit changes subsequently. This was where the coagulation of gases into the Sun and its surrounding planets, satellites, asteroids and comets occur. Or at least so young students of science are being taught.

QUESTION # 18. Is this model true? Really?

ISSUE # 19. In Chinese, there is a name for Mars, and it comes from very ancient times. It was, and still is "the fire star".

QUESTION # 19. But hasn't Mars always been just a speck of light in the heavens, and never closer than 30,000,000 miles to our planet? Why then would the ancient Chinese ever name Mars "the Fire Star"?

ISSUE # 20. Planets have energies, discrete energies. Energy is measured by the length of the "x" axis of an orbit, its long orbit diameter, the major axis. Maybe Mars, with its clobbered side, was out in the neighborhood of the asteroids, committing fragmentation, creating chaos.

QUESTION # 20. If Mars once was farther out in the neighborhood of the asteroids, wouldn't Mars in such a former orbit still have a similar amount of energy as it does in its modern orbit? See chapters 9, 10 and 11.

An Abundance Of Questions And A Paucity Of Answers

There are questions about Mars-Enlil"s presence on the day of Noah's Flood in ancient Sumerian literature. This writing was in wedge-shaped characters on clay tablets that still exist.

There are questions about Ares" presence during classical Greek times, the era of Homer and Hesiod. There are questions, also in classical Greek literature, about the sighting and scenario of its two little satellites.

There are questions about why ancient Mars would ever be characterized as a "fire star" in the ancient Chinese pictographic language.

There have been at least five fragmentation"s on the Roche Limit of various planets. Jupiter, Uranus and Neptune all have dark, rocky rings at their Roche Limits.

Saturn has a resplendent icy ring system at its Roche Limit. The indication is that the body that fragmented there was a small ice ball. Its icy debris, the ring system, is effervescing away slowly but steadily. That fragmentation was so recent that the ring system has not yet effervesced away. This indicates a recent dating, astronomically speaking, for another recent fragmentation on another Roche Limit, that of Saturn.

Therefore, why not do what seemingly is silly? Count the craters of Mars? They number in the high tens of thousands. And after all, haven't they been there for four billion years?

> From its superficial appearance, the southern hemisphere would seem to be generally older than the northern, and the larger craters, however they were formed, may date back 4,000 million years. [n7]

There are too many craters on Mars to count. There are tens of thousands, large, middle sized and small, and acquiring the appropriate photo mosaics is not that easy. Therefore, find a short cut. Why not count just those craters that have diameters of 20 miles or greater, and keep records of their locations?

Atlases of the Solar System are readily available in libraries and book stores. What data of interest data might a discrete count of the craters of Mars reveal?

The Project Of Counting The Larger Craters Of Mars

At the first impression, undertaking such a task on a planet that is, and always has been at least 35,000,000 miles from the Earth seems somewhat ridiculous. After all, haven't we all been taught that Mars has been just a tiny speck of light in the nocturnal heavens since the dawn of history? Maybe for 4,000 million years? Maybe, and maybe not.

This project was undertaken. The first stage of the project was to count just the craters 20 miles and larger. It was not difficult, though it was somewhat time consuming, the tally of this count, grid by grid, established that indeed there was a clobbered side, dense with craters and a serene side.

The second stage of the task was to ascertain where on Mars is the center of the Clobbered Hemisphere. Finding its center, it turned out, also was easy. Mars has a distinct rim where there is an abrupt fall off of crater density. It is easy to notice if one is thinking "planetary catastrophism." So far this "rim of craters", a drop off zone, has eluded astronomers thinking "gradualism". Once the rim's location is established, determining its center is a piece of cake.

The third part of the task was to ascertain the relationship, if any, between the center of the Hemisphere of Craters and the location of the giant Hellas Crater. Hellas Planitia, the largest crater in the Solar System, is close to the center of the Clobbered Hemisphere. Interesting.

In space mission research, Mars has arbitrarily been given latitudes, zero to 90°, both north and south of its equator. In this respect the latitudes of Mars are similar to the pattern of latitudes on the Earth.

Also, Mars has arbitrarily been given 360° of longitude. But they are designated from zero to 360° west. There are no "east longitudes" on Mars. On the Earth, longitudes are arranged 180° east and west of the Greenwich meridian to meet at the International Date Line. It is different on Mars where there are 360°, all being west of the arbitrary prime meridian. The 360th W. longitude meets, and becomes the zero longitude.

The next task was to divide up the surface into usable subdivisions, and count each subdivisions. The subdivisions were 10° grids, each having 10° of latitude by 10° of longitude. There were 648 such grids, a series of 18 for latitudes and a series of 36 for longitudes. 18 x 36 = 648. This is the number of individual grids of 10° on a side.

Next was the counting itself of craters, only those 20 miles in diameter or greater, grid by grid. There was no difference in the general distribution of larger craters versus smaller craters. So assessing the pattern of the larger craters will give the pattern of all of the Martian craters.

Counting the smaller craters would require hundreds of maps whereas counting the craters, 20 miles and larger, can be done on ten map pages of a good astronomical atlas, widely available at libraries. The count of craters of diameters of 20 miles or greater was made in each grid. The scale of our maps of the quadrants of Mars turned out to be 370 miles to the inch.

After counting and recording the craters, grid by grid, the grids were gathered and condensed into six regions. Region I was the northern region, 65° N. Lat. and north to the North Pole. Region II was a central region, 65° N. latitude to 65° S. latitude, and 0° long. to 90°. Region III was the same degrees latitude, but 90 to 180° longitude.

Region IV was the same degree latitude but 180 to 270° longitude. Region V was the same degrees latitude, but 270° to 360((or zero) longitude. Region VI was the southern region, 65° So. latitude and south to the South Pole.

Table I is a compilation of these crater counts, condensed into six regions. The results indeed are, to say the least, interesting.

The modern Southern Hemisphere is entirely in the Clobbered Hemisphere. It is heavily cratered, 355 craters. Its counterpart is entirely in the Serene Hemisphere. It had 25 craters, only 7% as many as its southern counterpart. There was a 93% versus 7% distribution for the two the polar regions. Would this same ratio hold for craters in the middle and lower latitudes of Mars? In fact, the same ratio did hold.

It is reasonable to assume that, before the fragmenting of Astra, both hemispheres of Mars had a random distribution of craters, 237 in each hemisphere.

Table I - Counts Of Craters On Mars - 20 Miles Or Larger In Diameter

GeographicalRegion	Clobbered Hemisphere	Serene Hemisphere
Region I - 65° N. Latitude to the North Pole	-	25
Region II - 65° N. Lat. to 65° S. Lat. 0° to 90° Long.	610	99
Region III - 65° N. Lat. To 65° S. Lat. 90° to 180° Long.	622	48
Region IV - 65° N. Lat. to 65° S. Lat. 180° to 270° Long.	655	48
Region V - 65° N. Lat. to 65° S. Lat. 270° to 360° Long.	826	55
Region VI - 65° S. Latitude to the South Pole	355	-
Total by Cratered Hemispheres	3,068 (93%)	237
Total Count of 20-Mile Craters on Entire Planet	3,305	

OVERLAID CRATERS

It is also reasonable to conclude that when the blizzard of fragments hit the Clobbered Hemisphere, 25% of the primordial craters in the Clobbered Hemisphere were masked out, and overlaid by newer craters. Thus, of the 237 primordial craters in the Clobbered Hemisphere, an estimated 59 craters once existed, but were overlaid by the blizzard of fragments. Hence, estimated 178 primordial craters remain in the Clobbered Hemisphere.

This means that the Clobbered Hemisphere received 2,890 craters, each 20 miles or greater in diameter, in one fifteen minute blizzard of fragments from Astra. (3,068 - 178 = 2,890). Before that, both the hemispheres of Mars were serene, and had a composite total of 474 20-mile craters.

From logic and Table I, the following information can be deduced..

1. 3,305 craters, each 20-miles in diameter or more, is the modern count for the surface of Mars.

2. 59 craters were, but are no more, having been masked and overlain by Astra craters.

3. 178 primordial 20-mile craters remain are in the Clobbered Hemisphere. The rest, 2,890, are from the blizzard of fragments of Astra.

4. 237 craters, all primordial, all 20 miles or more in diameter, are in the Seren Hemisphere of Mars.

5. 2,890 of the 3,068 twenty-mile craters in the Clobbered Hemisphere, are from the fragmentation Astra.

6. "0" - zero fragments of Astra, made craters in the Serene Hemisphere of Mars, as it was the backside of Mars, facing away from Astra.

7. 87.4% of the existing craters on Mars are from fragments of Astra (2,890 of 3,305). All 2,890 are in the Clobbered Hemisphere.

8. 12.6% of the existing craters on Mars are original craters, before Astra's fragmentation. Of these, an estimated 178 are in the Clobbered Hemisphere and 237 are in the Serene Hemisphere.

9. The distribution of the original, primordial craters on Mars has been random, and random in both hemispheres.

10. The distribution of craters from the fragments of Astra that clobbered only one hemisphere of Mars, is random in the Clobbered Hemisphere.

What does all this mean? It means the following.

A. If the tenth planet, Astra, once in the Solar System, wasn't already apparent from the spread of asteroids, it would have to be invented because of the spread of craters on Mars.

B. When Astra fragmented, Mars was there, and provided gravity and the Roche Limit for the crisis event.

C. The perihelion"s of the asteroids indicate the distance from the Sun when Mars and Astra were on a collision course. This distance will be fine tuned in Chapter 5. But that distance was over 200,000,000 miles from the Sun, but was less than 250,000,000 miles. THUS, WHEN ASTRA FRAGMENTED, MARS WAS NOT EVEN WITHIN 50,000,000 MILES OF ITS MODERN ORBIT.

D. What was Mars doing out there? It was riding its catastrophic orbit, which must have had an orbit eccentricity of somewhere between .50 and .65. Its orbit was more like Halley's Comet than the modern orbit of Mars, although it didn't go out as far as Jupiter. (For how far out, see Chapters 9, 10 and 11).

E. How far inward, then, did that catastrophic third orbit of Mars roam? If it roamed outward farther, the law of the conservation on energy requires that it also roamed farther inward.

Did Mars in its old orbit roam inward as far as 92,250,000 miles? If so, did it also roam inward as far as 65,000,000 miles, where it would bother Venus as well?

Figure 1- The Distance Between Astra and Mars at Fragmentation

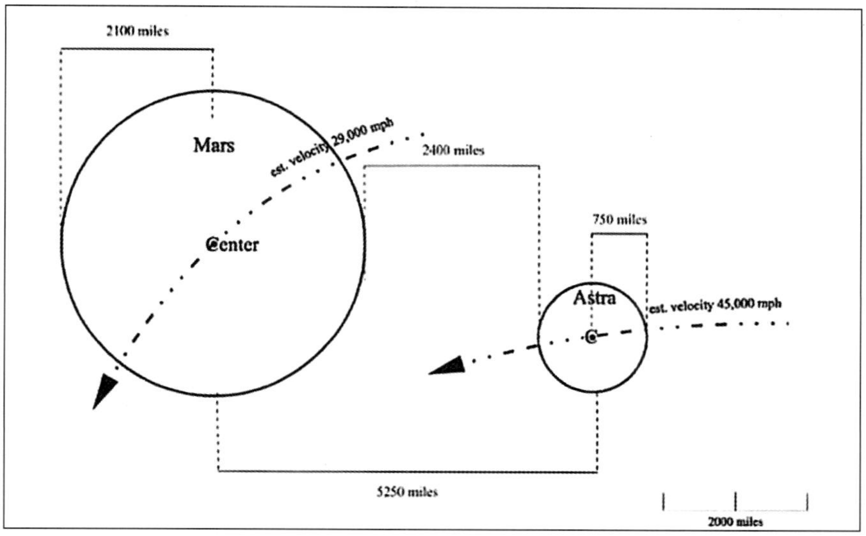

Figure 2 -Minutes After Astra's Fragmentation

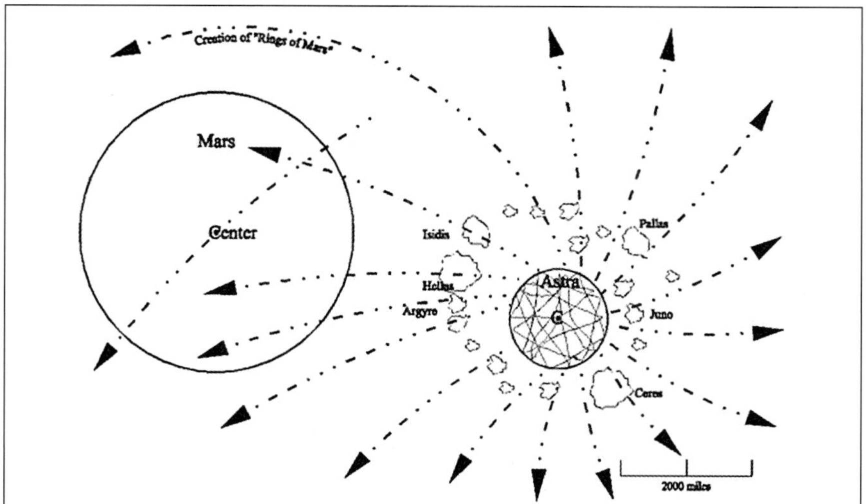

Two Minutes After Astra's Fragmentation

Figure 3 -The Battered Versus the Serene Hemispheres of Mars

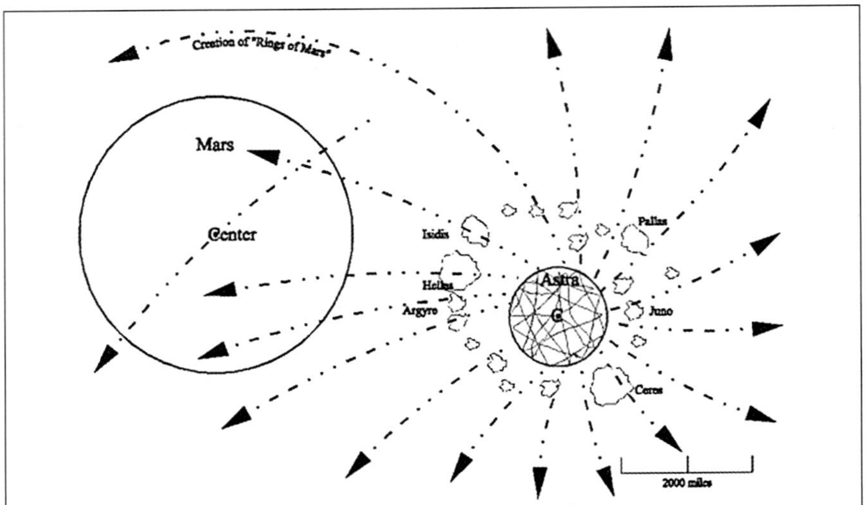

Two Minutes After Astra's Fragmentation

Evidence # 1 The Crater Distribution On Mars

Figure 1 illustrates the close up geometry of Mars and Astra as Astra was about to fragment. Figure 2 illustrates the same close up geometry of Mars and Astra just minutes after Astra did fragment. Figure 3 illustrates today's disproportionate distribution of Craters on the surface of the two hemispheres of Mars. It also represents the "Rim of Craters" on Mars where the density drops off dramatically.

Figure 9 in chapter 5 illustrates the orbits and the geometry of Mars and Astra from a great distance, as Astra overtook Mars in space, almost half way out to Jupiter's orbit. The orbital calculations and analysis of this team indicates that Astra was past its perihelion, when it overtook Mars. And old Mars, overtaken by Astra, was just past its ancient aphelion, out somewhere between 200,000,000 and 250,000,000 miles from the Sun. (The specific distance is contained in Table XI).

Figures 3, 4 and 5 summarize the crater distribution data on the surface of Mars. The data (not the figures) are potent evidence of a fragmentation of a small planet on the Roche Limit of Mars. That was when the Solar System had ten planets and no asteroids.

Evidence # 2 The Rim Of The Hemisphere Of Craters

There is an abrupt edge, or "rim" for a dramatic drop off of the density of craters on Mars. This rim is where "the buckshot ends". It is where the red planet's Serene Hemisphere begins. It is illustrated in Figure 3.

This rim is obvious to anyone who is thinking "fragmentation" on the Roche limit of Mars. So far, astronomers who fail to think "planetary catastrophism", also have failed to "see the obvious".

The rim rises farthest to modern north on Mars in its northeast quadrant, at latitude 48 N. and at longitude 320 W. From there, the Rim of Craters arches downward, or curves southwesterly toward the most southerly edge of the rim of craters.

The southern extremity of the rim, by modern coordinates, is at latitude 42 S., and at longitude 110 W. The "Rim of Craters" is not hard to identify if it is expected, or anticipated. It is there as it ought to be if Mars experienced a sudden, intense, 15-minute blizzard of fragments bombarding it on one side only.

This leaves everything in the modern Northern Hemisphere, above latitude 48 N., as being in the Serene Hemisphere. In the high northern latitudes of Mars, above latitude 65 N., a crater count was a made; for craters over 20 miles in diameter, it was a sparse 25, all presumably primordial craters.

In contrast, the entire Southern Hemisphere higher than the 42nd latitude S. is in the Clobbered Hemisphere of Craters. In the high latitudes of Mars, south of the 65th latitude S., the crater count (of craters 20 miles in diameter or more) is 355. Some of these, less than a dozen, could be primordial craters. The rest, almost 350 of them, are scars of Astra's fragmenting on the Roche Limit of Mars.

Statistically in these two zones, the cratering of the southern latitudes has been 7% in the high northern modern latitudes, compared to 93% in the high southern latitudes. That is the percentages of 380 divided by 25 and 355. This proportion favoring the modern Southern Hemisphere is 14 times as dense in cratering.

This disparity of crater distribution on Mars has led a typical astronomical editor to casually remark as follows:

In general terms the southern is the older of the two, this being indicated by the very great density of craters present ... The northern hemisphere is at a much lower elevation and is less densely cratered. [n8]

That assessment is true. Another astronomer remarks, equally casually, that the Southern Hemisphere is more densely cratered "for whatever reason". A sophomore in planetary catastrophism can improve easily on the depth of those insights.

Evidence # 3 The Gigantic Hellas Crater

In astronomical parlance, the Hellas Crater is termed "Hellas Planitia." The Hellas Crater is the biggest crater in this Solar System despite its location on such a small planet. Its longitudinal diameter extends from latitudes 29 to 54. On Mars there is 36.75 miles to a degree. This makes diameter of the Hellas Crater as 920 miles wide.

The latitudinal diameter of the Hellas Crater extends from longitude 272 to 312. Which makes it 980 miles long. The orientation of this shorter diameter suggests that the giant Hellas fragment which hit Mars here had one diameter somewhat longer than the other. That is normal for asteroid fragments.

Mars is a small planet with a surface area of 56,000,000 sq. miles. That is 28% of the area of the Earth's surface. Yet the Hellas Crater has a surface area of 710,000 sq. miles. Hellas Planitia covers 1.27% of the surface of Mars.

The Hellas Crater has as much area as Alaska, with Washington and half of Oregon thrown in. The "H" Crater is 2.66 times bigger in area than Texas. This gigantic crater, on such a small planet, has as much area as 23 of the 26 states east of the Mississippi River combined; all the eastern states except Illinois, Michigan and Wisconsin.

In Europe, the area of the Hellas Crater is equal to the combined areas of Great Britain, Ireland, France, Belgium, Luxembourg, the Netherlands, Switzerland, Germany, Denmark, Switzerland, Austria, Italy, Monaco, San Marino, Liechtenstein and half of Hungary.

It is estimated that the Hellas Fragment was some 600 to 625 miles in diameter. It clobbered the facing side of Mars at a velocity of 25,000 mph, or, 420 miles per minute, or 7 miles per second.

It so happens that the largest of the asteroid, Ceres, also is 625 miles in diameter. The Hellas fragment and the Ceres fragment may well have been "twins" in size.

Commensurability Of Sizes - Asteroids And Mars Craters

Pallas is the second largest asteroid, diameter 380 miles. Isidis is the second largest crater on Mars, diameter 650 miles. If, as an asteroid, the Isidis Fragment were 380 miles in diameter, and if an asteroid like Pallas made a crater 75% larger that the its diameter, then Pallas and the Isidis Fragment may have been similar in size. It is suspected that asteroids of this size and velocity made craters 75% larger in diameter than the fragment size.

The third largest crater on Mars is Argyre, diameter 400 miles. The fragment that created Argyre Planitia may have been as big as Euphrosyne, diameter 229 miles.

In comparing the sizes of the asteroids to the craters on the Clobbered Hemisphere, three things need to be realized. First, all 15 of the largest craters of Mars are in its battered, Clobbered Hemisphere. Second, about twice as many asteroids missed Mars as hit it.

Third, it is estimated that an asteroid hitting the Clobbered Hemisphere at such a high velocity will make a crater 75% larger than the fragment's diameter. With these in mind, Table II gives a comparison of the sizes of asteroids and the sizes of craters in the Clobbered Hemisphere of Mars.

Table II - A Comparison Of Diameters - Asteroids To Mars Craters

ASTEROID	DIAMETER		MARS CRATER	DIAMETER
1. Ceres	622 miles		1. Hellas	990 miles
2. Pallas	377		2. Isidis	684
3. Vesta	334		3. Argyre	481
4. Hygeia	279		4. Cassini	291
5. Euphrosyne	229		5. Schiaparelli	282
6. Interamnia	217		6. Antoniadi	222
7. Davida	200		7. Schroeter	185
8. Cybele	192		8. Name not known	175
9. Europa	179		9. Herschel	158
10. Patienta	171		10. Kepler	150
11. Eunomia	169		11. Newcombe	144
12. Juno	155		12. Secchi	139
13. Psyche	153		13. Schmidt	133
14. Doris	155		14. Flaugergues	132
15. Undina	155		15. Kaiser	130

As was mentioned above, it is thought that fragments of these sizes made crater diameters on Mars 75% larger than their own diameters. If this is so, and if two thirds of the asteroids missed Mars, and one third hit the Clobbered Hemisphere, then these two lists are commensurate. The largest asteroids and the largest of the Astra fragments hitting Mars were of similar dimensions.

The fragments of Astra quickly spread out into a vast spray, a great cloud of asteroids that was somewhat larger in diameter in 12 minutes than was the diameter of Mars itself, at 4,212 miles.

Discovery consists of seeing what everybody has seen and thinking what nobody has thought.- Albert Szent-Gyorgi (discoverer of Vitamin C)

Conclusion

In Volume I, a series of 18 new insights were presented in astronomy/cosmology involving the formation of this Solar System. They were insights based on the paradigm of planetary catastrophism.

It was concluded that the planets were delivered to the Sun on the same plane, with their satellite systems intact, and with their spin rates already operational, craters in position, etc. Evidence was given in eighteen categories for this and that the Sun captured these planets "recently" as an astronomer assess time.

Story 1 in this series contains the evidence and the logic that THE ROCHE LIMIT OF NEARBY MARS WAS THE SPECIFIC AGENT CAUSING THE FRAGMENTING OF A FORMER TENTH PLANET, ASTRA, INTO ASTEROIDS. The two planets were on a collision course, with Astra overtaking the red planet. The timing was thousands of years ago, not millions.

Story 2. One of the results of Astra's fragmentation was that some 30% to 33% of its fragments bombarded one side of Mars, MAKING THAT FACE OF MARS AMONG THE MOST CLOBBERED HEMISPHERES IN THE SOLAR SYSTEM.

This bombarded hemisphere of Mars was subjected to a 15-minute blizzard of fragments of all sizes up to 600 miles in diameter. They numbered into the high tens of thousands, and hit the surface of one side of Mars at a velocity estimated at 25,000 mph. A foundation for this velocity estimate is laid in a subsequent chapter.

Story 3 is the "RIM Of CRATERS" on Mars. This rim describes the edge of the Clobbered Hemisphere, where there is a sharp, dramatic drop off of crater density. This rim fully encircles the red planet, but not along modern lines of latitudes or longitudes. The Hemisphere of Craters is predominantly in the modern Southern Hemisphere of Mars, but yet reaches to the 48th latitude N. in the modern Northern Hemisphere of Mars at one place.

Story 4 is that the CENTER OF THE CLOBBERED HEMISPHERE IS AT LATITUDE 45° S. AND LONGITUDE 320° W. The western edge of the Hellas Crater is only 220 miles east of this central location. If the Battered Hemisphere is viewed as a target, the Hellas Crater is in the bulls eye zone. It was over this surface location, some 2,500 miles high, where Astra penetrated the red planet's Roche Limit, and fragmented into tens of thousands of fragments.

The sizes of the asteroids and the sizes of the fragments that created the craters on Mars are commensurate within the standard that some 65% or slightly more of the fragments of Astra missed Mars, while only 30% or a little more hit the red planet's Hemisphere of Craters.

Story 5 is that the biggest little bang of this Solar System did not occur near Jupiter, Saturn, Uranus or Neptune. Judging strictly by the size of the asteroids it produced, THE BIGGEST "LITTLE BANG" HAPPENED JUST NEXT DOOR, ON THE ROCHE LIMIT OF OUR NEIGHBORING PLANET, MARS.

The diameter of Astra is estimated at 1,500 miles, plus or minus 100. It was about 26% smaller than the Moon, and some 15% larger than Pluto.

When Astra fragmented on the Roche Limit of Mars, Mars was between 200,000,000 and 250,000,000 miles from the Sun. Its present orbit at its furthest, is 155,000,000 miles from the Sun. Therefore, something radical has changed the energy of Mars, and something radical has changed its orbit eccentricity. What was that "something?" (It happened during the era of ancient literate man).

The ancient Chinese indicted Mars as "the fire star." The ancient Sumerians indicted Enlil (Mars) as the cause of Noah's Flood. The Greeks of classical times including Homer indicted Ares as the "bane of mortals." ("Bane" means murderous, or highly destructive).

Talmudic accounts, apparently dating to the 6th century B.C.E., indicted Sammael was an evil archangel, and also indicated that it was the comptroller, or "choreographer" of Mars. Other better angels were the comptrollers of the Sun, Moon, Mercury, Venus, Jupiter and Saturn. Archangels were viewed as messengers of the Lord, but their scale sometimes was planetary.

It is beginning to look as if their assessments of the cosmos were correct. Gradualists still hold that Mars has never come closer to the Earth than 30,000,000 miles at any time over the last 4+ billion years. Ancient sources indicate it came close to our planet fairly often, in fact once or twice a century. It is beginning to look as if the scientific score is Ancient Catastrophists, 1; 20th Century Scientific Gradualists, 0.

CHAPTER 2

The Tharsis Bulge of Mars

> *It is the greatest discovery in method which science has made that the apparently trivial, the merely curious, may be clues to an understanding of the deepest principles of nature.*
>
> J. J. Thompson, 1936,
> Recollections and Reflections, London, Macmillan

Introduction

Thus far, three categories of evidence have been cited concerning the former existence of a tenth planet, a small Pluto-sized body. Pluto's diameter is listed at 1,375 miles; Astra's diameter may have been 200 miles greater. That tiny planet was on a collision course with Mars, and fragmented on Mars' Roche Limit. Its fragments created a hemisphere of craters, accounting for more than 85% of all of the Martian craters. But that fragmentation created something more.

The three categories of evidence on the fragmentation of Astra discussed to this point are (a) the gross imbalance in the uneven distribution of craters on Mars, (b) the giant Hellas Crater in the bulls eye zone of the Clobbered Hemisphere, and (c) the rim where the density of craters drops off markedly and obviously.

That tiny tenth former planet needs a name. In astronomy, by tradition, discoverers are given the privilege of naming their discovery(ies) The name "Astra" is hereby nominated; it is cognate with asteroids, one of the main products of this fragmenting event. Also traditionally, the name of a new planet has come from Greek celestial mythology. Uranus, Neptune, Pluto.

In Chapter 6 there is a discussion of Greek mythology and the role therein of the Greek deity, Astra. Information there about Astra and her sojourn in the heavens will add a second reason to nominate the name "Astra."

The fourth in this series of clues isn't even in the Clobbered Hemisphere, where the Hellas Fragment hit. It is more or less opposite to the giant Hellas Crater, in the middle of the Serene Hemisphere. It is (d) the Tharsis Bulge.

The Tharsis Bulge of Mars geologically is considered to be a "shield" It is a bulge that is assessed by Moore and Hunt to be 5,000 km. (over 3,100 miles) broad. This diameter, if Moore and Hunt are correct, represents 23% of the circumference of Mars and 46% of the circumference of the Serene Hemisphere.

The height of the Tharsis Bulge is 23,000 feet above the surrounding plain. As it is with the Hellas Crater, the largest crater in this Solar System, so it is with the Tharsis Bulge. It, too, is the biggest bulge among the planets that have solid surfaces.

Evidence # 4 - The Tharsis Bulge

The giant Hellas Crater is located near to the center of the "Clobbered Hemisphere", or "Battered and Blasted Hemisphere." Apparently it was caused by a fragment almost the size of the largest asteroid, Ceres, which is 625 miles in diameter. The Hellas fragment created a huge crater, slightly elliptical, with a short diameter of 920 miles and a long diameter of 980 miles. In addition, the Hellas Fragment created a general indentation in the Clobbered Hemisphere.

Since Hellas Planitia is near the center of the Clobbered Hemisphere, this huge fragment must have hit Mars a direct blow, not a glancing blow. A foundation will be laid in a coming chapter to establish its velocity, 25,000 mph. This is 420 mp minute, and 7 mph second.

The crust of Mars is estimated at 20 miles thick, an opinion for which a foundation will be laid shortly. Thus the Hellas fragment penetrated entirely through the front crust of Mars, at a high velocity, and plowed into and almost through the fluid magma of Mars. It may be that the back crust, with a better angle of curvature, was what stopped the Hellas Fragment from going entirely through Mars.

Evidence will appear in a coming chapter indicating that both Astra and its fragments had spin rates, rapid spin rates, in addition to an orbital velocity of 25,000 mph. Depending on its diameters, the Hellas Fragment may have contained 25 to 50 million cubic miles of rocky material.

Once beyond the crust and into the hot magma soup of Mars, the Hellas Fragment created sudden, immense pressure waves. Pressure waves were in the front of the plunging Hellas. Shear waves were also involved; they were perpendicular to the direction of the plunging Hellas. Thus the entire inside of the Serene Hemisphere suddenly suffered immense pressures. It bulged, apparently where the crust was the thinnest and the weakest.

The Tharsis Bulge is located within 30° of being opposite to the Hellas Crater. It probably represents the region in the Serene Hemisphere where the crust was thinnest and weakest. It was where the crust was first to yield to the new sudden immense pressure. The inner side of the crust of the Serene Hemisphere yielded up 23,000 feet, over 4 miles, which is the height of the bulge.

Thus, the Hellas Fragment was unable to pass clear through Mars, but it was able to do ample damage; it suddenly thrust up the broad Tharsis Bulge. This bulge began to rise within 90 minutes of when Astra fragmented.

Moore and Hunt indicate the breadth of Tharsis is 3,100 miles, and its height is 23,000 feet above the surrounding terrain. (There is no mean sea level on Mars.) They describe the Tharsis Bulge as follows.

Tharsis Region. This region includes the most prominent volcanoes on Mars. The contours outline the Syria Rise, an enormous bulge in the Martian crust 5,000 km [3,100 miles] across and 7 km high [4.35 miles or 23,000 feet]. This bulge is the site of Ascraeus Mons, Pavonis Mons and Arsia Mons. [n1]

Mars is a small planet compared to the Earth. Its mass is only 10.7% of our planet's mass. Measuring from the Earth's sea level, the elevation of Tharsis exceeds all the peaks of the mighty Andes including Aconcagua, at 22,834 feet. The elevation of the Tharsis Bulge also exceeds all except the highest 22 peaks of the mighty Himalayas.

Were Tharsis on the Earth, its top would be in rarefied air. At 23,000 feet, three-fourths of the Earth's atmosphere is below. At 23,000 feet, barometric pressure is only 25% that of sea level. But Mars has no sea level and it has less than 1%, as much atmosphere as the Earth.

The Tharsis Bulge dominates the physical geography of the Serene Hemisphere of Mars. On it are the six largest volcanoes in the Solar System. The next six largest are in the region of the Elysium Bulge, the second bulge in the red planet's Serene Hemisphere.

The Probable Thickness Of The Crust Of Mars

Figure 4 illustrates the plunging Hellas Fragment. It encountered, successively, first the front crust of Mars, second, its hot magma interior, and perhaps third, its back crust. Crustal thickness on Mars is at issue in understanding both this bulge, and the giant volcanoes thereon. Its thickness indicates the resistance, of the front crust of the Hemisphere of the Clobbered Crust, to plunging large fragments like Hellas, Isidis, Argyre and also the smaller fragments. And it may indicate the resistance of the back crust of the Serene Hemisphere, denying Hellas a passage entirely through the red planet.

The crust of the Earth is considered to average ten to twelve miles in thickness. It is also considered to be elastic and flexible up to a point. It has an elastic-plastic threshold beyond which it will tear, but that threshold is never approached in this modern serene age.

The crust of Mars is considered to be thicker and more rigid, and much less elastic for several reasons. First, Mars is made of lighter materials than is the Earth. Earth's density is 5.52, Mars' density is 3.93 (and water is 1.0) The density of the crust of Mars is only 72% of the Earth's crust.

On Mars, as in the Earth's crust, temperatures rise going down vertically. But in going down, to where the temperatures can melt crust, temperatures rise only 70% as fast on Mars. Temperatures must rise to 3000° F. to liquefy the crustal materials such as silica and alumina.

Second, Mars is farther from the Sun by 52% than is the Earth. Hence its crust is colder at the beginning. Temperatures on the cold surface of Mars average -90° F. On a square mile to square mile basis, the surface of Mars receives only 43% as much radiation from the Sun as does the Earth. During the 24-hour 37-minute night, surface temperatures of -250° F. are common. So the surface of Mars has an average temperature of about 130° F. colder than the crust of our planet.

Third, Mars has almost no atmosphere. The Earth has an atmosphere that absorbs and retains a significant part of the heat that the Earth radiates to space. It is called captured radiation. The atmosphere of Mars captures very little radiation from the surface of Mars.

Fourth, Mars has only 55 million square miles of surface; the Earth has 196 million. Thus, Mars has 28% as much surface as does the Earth, but it has only 11% as much mass as the Earth. Smaller planets with more surface per unit of mass radiate heat back into space more efficiently than do larger planets like the Earth and Venus.

This is why surface temperatures on Mars rise and plunge so rapidly, as much as 300° F. in a 24-hour period. At night, temperatures on Mars can drop 20+° F. per hour. Thus very cold temperatures penetrate to considerable depths in the crust of Mars. The crust of Mars is estimated to be 20 miles thick.

Fifth, the crust of Mars has a smaller radius of curvature than does the Earth's crust. Beyond the accelerated radiation factor, this gives the crust of Mars an added stiffness factor - a rigid inflexibility - that is not characteristic of the Earth's crust.

Thus Mars has both a thicker crust and a more rigid crust. Both features will make it more difficult for fragments of Astra to penetrate its crust than if they hit the surface of the Earth. However, it is clear that at least three fragments of Astra penetrated into the mantle of Mars, Hellas, Isidis and Argyre. Another ten or fifteen may have also. But most of the fragments under 100 miles in diameter didn't penetrate, even when hitting its crust at 25,000 mph.

The Location Of The Tharsis Bulge

The heartland of the Tharsis Bulge is located between 101 and 125° W. longitude. Our measurement is for breadth about 900+ miles. Our measurement for its length is from 16° N. to 12° S., making it over 1,000 miles. This makes its heartland area about 650,000 miles. For whatever reason, our measurements for the heartland of Tharsis are more conservative than those of Moore and Hunt for the entire bulge.

The evidence emerges that the Hellas fragment hit the crust of Mars a direct blow, from almost vertical. It passed into the internal magma of Mars, creating enormous pressure waves and shear waves. The Hellas Fragment did not pass out through the other side of the crust of the Serene Hemisphere. But the angle of the hit, its velocity, and the Hellas Fragment did cause sudden, immense internal distress, resulting in a huge pair of bulges in the opposite hemisphere.

Apparently the Hellas Fragment plunged into and through the crust and continued to plunge onward, rotating all the way, for up to three thousand miles through the magma of Mars. To describe the internal distress of Mars, it can only be said that it was beyond chaos cubed.

In the case of the Earth's atmosphere, shock waves have a velocity of about 750 mph. Sometimes they are called sonic booms. However velocities of pressure waves in the oceans are different; in water they are four times as fast, 3,000 mph. Presumably, pressure waves in the magma of Mars were around 3,000 mph.

If so, the fragments of Astra hit the front crust some 11 or 12 minutes after Astra fragmented. The pressure waves, at an estimated 3,000 mph, required another 80 or 85 minutes to arrive at the Tharsis Bulge. Thus, the Tharsis Bulge began to rise, with suddenness, about 100 minutes after Astra fragmented.

The crust and the magma of Mars first had to slow down the speeding, plunging "H" fragment from 25,000 mph. Then its magma had to deal with the pressure waves and shear waves which the "H" fragment produced.

Simultaneously there were at least two other fragments that penetrated through the Martian crust, Isidis and Argyre. In the vicinity opposite the Isidis Crater is the second bulge of Mars, known as the "Elysium Bulge." It also contains a spread of volcanoes, huge by Earth's standards, but small by the standards of the volcanoes on Tharsis.

Thus the internal distress within Mars after Astra fragmented was "chaos cubed" It is beyond mathematical analysis. It was a wild and woolly day for the both the crust and the magma of Mars.

The Serene Hemisphere And The Elysium Bulge

The Isidis Crater is not far from the Rim of Craters on Mars in its Northeast Quadrant. Isidis is the second largest crater on Mars. This crater has a diameter of 450 miles and an area of 175,000 miles. The area of the Isidis' crater is equal to that of California, with Maryland and New Jersey added in for good measure. This is on a much smaller (and colder) planet.

Not far from the Isidis crater, and also in the Serene Hemisphere, is the Elysium Bulge on Mars. Like Isidis, the second-largest crater, Elysium is the second largest bulge on the red planet.

The Argyre Crater, the third largest, is about 300 miles in diameter, and contains some 75,000 sq. miles. Its area is equal to the state of South Dakota. Or, on the Eastern Seaboard, it is equal to the combined area of Delaware, the District of Columbia, Long Island, Maryland, New Jersey, Virginia and half of West Virginia. Where these two bulges are is an indication of where the thinnest parts of the crust of Mars are.

Figure 4 - - The Plunges of the Three Largest Fragments of Astra (Hellas, Isidis and Argyre)

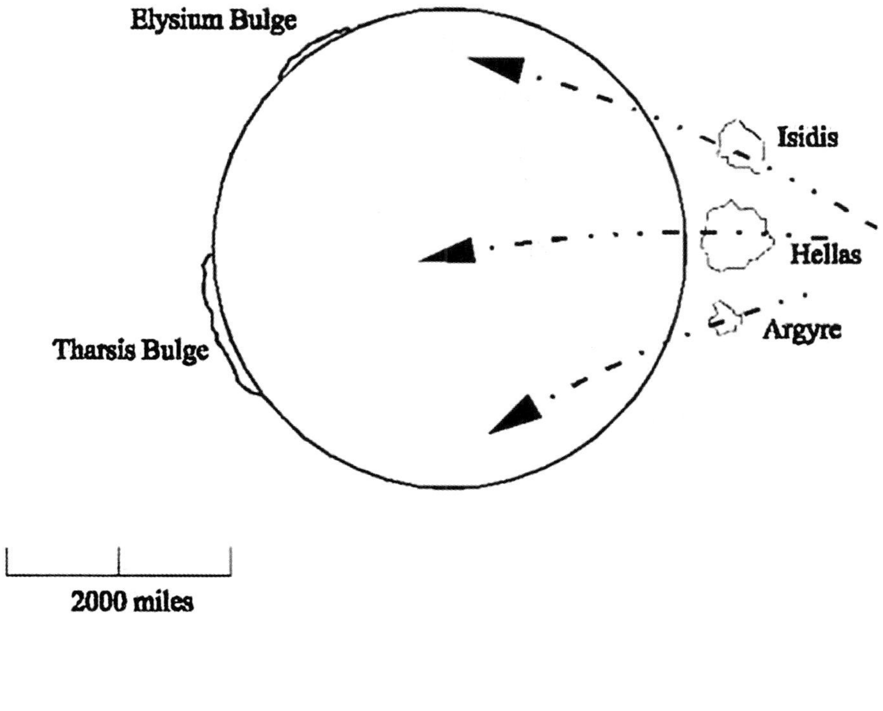

The Plunges of the Three Largest Fragments of Astra (Hellas, Isidis and Argyre)

When Astra fragmented, and planetary catastrophism came to the planet Mars, it came in a fast, fierce, furious flurry. Similarly, when planetary catastrophism came to the planet Earth, it also came fast, fierce and furious. The calculations are that the fragments of Astra approached Mars at 25,000 mph. By comparison it calculates that Mars approached the Earth at a velocity differential of 30,000 mph.

The following is a citation about quaking, shaking and crustal deformation on another planet, the Earth, within the recorded memory of mankind. It was about 3,700 years ago, not 3,700 millions of years ago.

Suddenly He moves the mountains, overturning them in his anger. He shakes the Earth to its foundations. The Sun won't rise, if he commands it so. Only He has stretched the heavens out and stalked along the seas. He made the [constellations] Bear, Orion and the Pleiades, and the constellations of the southern Zodiac. Job 9:5-9, Living Bible.

Conclusion

Each story in this skyscraper of catastrophic cosmology indicates a new understanding for a puzzling feature in this Solar System. Story 6 is the story of THE SUDDEN UPLIFTS OF THE THARSIS BULGE AND THE ELYSIUM BULGE ON MARS. Rising within a matter of a few hours, it was in no sense a gradualistic uplift.

A study of pressure waves and shear waves suggests Tharsis was beginning to be uplifted approximately 85 minutes after the gigantic Hellas Fragment blasted through the crust of Mars on the opposite side.

Simultaneously the Elysium Bulge was uplifted. Both of these regions of uplift probably reflect where the crust of Mars in the Serene Hemisphere was thinnest and weakest. They could least endure internal pressure and shear waves. The fact that both of these regions, and only these two regions also contain the volcanoes of Mars underlines that here, the crust of Mars was and is the thinnest and weakest.

The "H" Fragment, still rotating, first plunged through the 20-mile thick, rigid crust of Mars, which reduced its velocity somewhat, but not enough. Next its velocity was reduced by plunging through the 4,000 miles of the molten magma of Mars.

Finally the "H" Fragment came to a standstill by resistance of the underside of the crust of the Serene Hemisphere on the opposite side. Even then it was still rotating a bit. The suddenness, the immensity and the velocities of the two fragments were what created the two bulges. The dating of the fragmenting of Astra is touched upon several times in the following chapters. It was not 3.0 to 3.5 billion years ago.

Under the circumstances, it is not surprising that bulges developed suddenly in the

Serene Hemisphere. Crustal bulging was one major method, but not the only major method in the process of relief for sudden, internal distress of Mars, an internal distress that suddenly gripped its inward regions. Volcanism also helped (see Chapter 4.) Rifting helped even more (see Chapter 3).

A. B. Guthrie of Choteau, Montana, a famous author of Western Americana, described the crest of the Rocky Mountains or the Continental Divide, as being "high, wide and handsome." Much higher, wider, and more handsome are the Tharsis and the Isidis Bulges on Mars.

With story 6, the reader now is 33% of the way to the penthouse. Each story provides a new insight for cosmology, the history of the Solar System, and each story provides an increasingly broad view of the human history on this planet.

CHAPTER 3

Mars Puts On A Little Weight

Man might as well project a voyage to the moon as attempt to employ steam navigation across the stormy Atlantic Ocean.
Dr. Dionysius Lardner, 1838

Introduction - The Estimated Diameter Of Astra

Astra's Size. The diameter, mass and density of Astra at best can only be estimated. One method of estimating its former diameter is to review and assess the diameters of its largest asteroid components. These include (1) the fragments that became asteroids, over 65%, and (2) the fragments that hit Mars, over 30%. It was given in a previous chapter that the ten largest asteroids have the following diameters in miles:

Table III - Diameters of the ten largest asteroids

Ceres	624 miles		Interamnia	217 miles
Pallas	378		Davida	201
Vesta	34		Cybele	192
Hygeia	280		Europa	180
Euphrosyne	217		Patienta	172

Astra exploded into fragments, tens of thousands of which hit and impacted the Clobbered Hemisphere of Mars. Our estimate for the size of these fragments is 55% to 59% of their huge crater diameters. Therefore the estimates for the diameters of the ten largest of these original fragments are as follows.

Table IV - Diameters of the ten largest original fragments

Hellas	565 miles		Antoniadi	127 miles
Isidis	390		Schroeter	105
Argyre	274		Name Not Known	100
Cassini	166		Herschel	90
Schiaparelli	161		Kepler	86

The density of Mars is 3.93 and the Moon's density is 3.34. The Earth's density is 5.52 (water - 1.0). Judging by the composition of the asteroids, Astra's density seems to have been much like the density of Mars, 3.93. If so, this information can help estimate the mass of Astra. Astra with a diameter of 1,550 to 1,600 miles, was 4% to 5% of the mass of Mars, approximately one-twentieth.

By comparison, the Moon is 11% of the mass of Mars, one ninth. Thus it is estimated that Astra was almost half of the mass of the Moon, and slightly denser. If so, a body 45% as massive as the Moon fragmented on the Martian Roche Limit. The Martian Roche Limit, the place of fragmentation, was 5,200 miles from the center of Mars, but only 3,100 miles from its nearest surface.

Evidence # 5 - Crustal Expansion

How much of the mass of Astra became asteroids? And how many asteroids have been counted? The current asteroid count exceeds 5,000 and increases yearly. How much of Astra's mass was ingested by Mars because of hitting the Clobbered Hemisphere of Mars? Only estimates can be made.

This estimate is that between 30% and 33% of the fragments of Astra hit Mars in its Clobbered Hemisphere. Between 65% and 68% missed Mars and began to orbit the Sun. Also, a small percent, perhaps almost 2%, missed Mars but began to orbit around Mars instead of the Sun. Evidence of this, presented in Chapter 6, is the foundation of an ancient ring system of Mars.

If 30% of the fragments of Astra hit Mars, containing 30% of the mass of Astra, and if Astra was 5% of the mass of Mars, then it follows that Mars gained some significant weight. It gained perhaps 1.5% in mass by absorbing fragments of Astra.

A gain of 1.5% for a 200 lb. man, loosely clothed, is only three pounds. It would be marginally noticeable. But how would a sudden gain of 1.5% in mass affect the tight-fitting shell, or crust of Mars? A shell 20 miles thick on a diameter of 4,200 miles.

Would it bulge? Would it crack and split? Would such a split be perpendicular to the sub crustal pressure, allowing for expansion? The sudden ingestion of the Hellas Fragment caused a bulge. The sudden ingestion of the Isidis fragment created another bulge. But that was merely the beginning.

Rift Systems And Their Genesis On The Earth

Rifts are large, transverse, lineal "tears" or rips across the tight-fitting crust of a planet. One cause of a rift can be that a planet gains weight (mass). Another cause can be that its weight does not increase, but its weight shifts location - a relocation of an equatorial bulge for instance. Rifts can occur in a crust if there is either kind of shift in weight.

The Earth has had such a shift in weight, a relocated equatorial bulge. The cause of the Earth's Great African Rift Valley has been a change in distribution of its mass. It was due to a radical relocation of the spin axis, and a consequent radical relocation of the equator, and consequently of its equatorial bulge.

The Earth has a polar diameter of 7,900 miles but its equatorial diameter is 27 miles greater, at 7,927 miles. The Earth's spin rate is once in 1436 minutes. This spin rate creates its 27-mile equatorial bulge. It also creates two polar flat spots.

Mars rotates at a spin rate almost identical with Earth's, and also has an equatorial bulge. Its equatorial diameter is 4,212 miles. Its polar diameter is 4,170 miles. Its equatorial diameter is 42 miles greater than the polar diameter. It is an oblate spheroid. Its oblateness is .0093. Tiny Mars has a mass of .107 of the Earth, and a spin rate is once in 1477 minutes, only 3% slower than the Earth.

The Earth is a gyroscope in space. Were it to experience a close Mars flyby, the resulting torsion would cause its spin axis to shift its location. The poles would swerve and careen, and would shortly relocate. Its equatorial bulge also would relocate.

A foundation will be laid in Volume III that our North Pole on one occasion shifted at least 3,000 miles during the closest of all of the Mars flybys. That produced a gross relocation of the North Pole by 3,000 miles, as well as of the equatorial bulge and the two polar flat spots.

The shift of the Earth's equator required an expansion of the new equatorial bulge in a new belt across the Earth's middle. The response of the Earth's crust, like a garment, was to tear. The rip was perpendicular to the new equator. The Great African Rift Valley is that rift, a perpendicular rift caused by this radical (and recent) shift of the equator.

The Grand Canyon and the Copper Canyon also are earlier tears in our crust, earlier, also resulting from spin axis relocations. Several rifts have occurred in the Earth's crust above sea level. More have occurred below sea level. 72% of the Earth's surface is covered by oceans.

If a man wears a tight-fitting suit or pair of pants, when the cloth is stressed, it tears. It tears perpendicular to the force exerted. Overalls tear across the knees of pant legs when the cloth stretches perpendicularly, beyond the elastic-plastic threshold.

It is all very simple. The crust of the Earth tore across the equator, in a perpendicular manner, to accommodate the new, relocated bulge zone. In modern English it is "The Great African Rift Valley." It is the Earth's biggest rift system.

The Rift System And Its Genesis On Mars

The Great African Rift Valley is deep, wide and handsome. It is 5,000 miles long, over 200 miles wide in its central region, and is over 4,000 feet deep in places. It stretches from Syria, through the Red Sea and across the Central African highlands, to Mozambique and beyond, into and under the Indian Ocean. But what, for a rift valley, is "deep, wide and handsome?"

The Solar System's Biggest Rift System

In the case of the crust of Mars, its rift system was not triggered by a SHIFT of the location of weight, as has been the case with the Earth's crust. It was created by a SUDDEN INCREASE IN WEIGHT, OR MASS OF THE PLANET ITSELF. The gain in weight was due to the sudden ingesting and then digesting of the fragments of Astra. That increase in mass is estimated at 1.5%. If Mars were to increase in mass by 1.5% due to ingestion of fragments of Astra, its equator would expand by 0.5%, or 21 miles. Its polar diameter would increase by 20 miles. This means its old equatorial diameter was only 4,191 miles and it sold polar diameter was only 4,150 miles.

The crusts of Mars and the Earth are tight fitting, unlike a suit of clothes. Obviously with sudden stress and a sudden gain in weight, the old crust, a rigid shell, would experience pressures beyond what the elastic-plastic threshold could cope.

It has already been noted that the two bulges of Mars occurred in its Serene Hemisphere, where its crust apparently is weakest. All of the rather huge volcanoes of Mars also occur in its Serene Hemisphere, further suggesting the Serene Hemisphere has a thinner crust. Third, its Valles Marineris, the greatest rift valley in our Solar System, also occurs in its Serene Hemisphere. Chance? No. Region of its thinner crust? Yes.

It was on the underside of this hemisphere on which the pressure waves and shear waves of the Hellas Fragment were spent. They were caused by the penetrating asteroids, Hellas, Isidis and Argyre. In another chapter, a foundation will be laid that all three of these huge fragments were rotating as they hit at Mars' Clobbered Hemisphere at 25,000 mph.

HORSTS. In geology, large, block-like uplifts on our planet were named originally by German geologists in the 19th century. A block-like uplift such as those found prominently in the Balkans was named a "horst." A standard geological definition of a horst is:

Horst: An elongated, elevated block of crust forming a ridge or plateau, typically bounded by parallel outward-dipping normal faults.

GRABENS. Grabens are intimately related to, and are associated with horsts. Grabens are the valleys between horsts, defined as follows:

Graben: A down thrown block between two normal faults of parallel strike but converging dips; hence a tensional feature.

The greatest grabens and horsts in our Solar System are on the surface of the Serene Hemisphere of Mars - the Valles Marineris. When this giant rift system was first photographed some 30 years ago, astronomers and geologists were aghast when viewing the first photographs by Mariner 2 in August, 1963. Mariner 2 made a flyby of Mars at 22,000 miles and transmitted its pictorial data back to the Earth. Shortly thereafter, this gigantic "horst and graben formation" on Mars was named the "Valles Marineris."

The first viewers of the Valles Marineris were astounded astronomers. Next, and even more astonished, were the jumpy geologists, always needing to defend the dogma of gradualism - in the Solar System as well as on the crust of the Earth.

That has been a pattern that continues. First the astronomers find the evidence, and then the geologists justify, or attempt to justify the chaotic scenes of profound violence and scenes of sudden spasms of catastrophism. Three or four billion years ago, they are apt to say. On Mercury's surface. On Venus. On Mars. On Deimos and Phobos. On the satellites of Jupiter, rings of Saturn and Uranus, ragged, jagged satellites circling Uranus, etc.

Surprises in the cosmos have been so regular and so astounding to gradualists that it makes one wonder if their paradigm is in proper order. It is now over thirty years later since the first photos of the surface of Mars were received, including photos of the Valles Marineris. Its details still startle geologists.

It is not just the huge size of the rift valley. It is not just the size of some of the craters. It is not just the lop-sided distribution of Martian craters. It is not just the crustal bulges, with their five and ten-mile high volcanoes. It is not just evidence of spin axis shifts.

Equally astounding, if not more so, is the recognition of the great rift system of Mars. It is a new fact in a world where old gradualist dogma still dominates. Until this chapter, there has not been much of a theory as to why it exists or how, or when in its freshness, it came to be.

It is a giant crustal rift valley, 2,500 miles long, 4.5 miles deep (not 4,500 feet), and is wider than an eye on the ground can see. Its width in places is 300 miles. The Earth has nothing comparable. Though explanations for this striking feature to date have not been forthcoming, yet the explanation is fairly simple. It was a matter of either shifting the location of weight (as in the Earth's case) or a permanent gain in weight (1.5% as in the Martian case).

The Valles Marineris

Mars has a circumference of 13,200 miles. The Valles Marineris is as long as 19% of the circumference of Mars. It is a 2,500-mile long system of rifts, of grabens and horsts, emanating eastward out of the eastern edge of the Tharsis Bulge. Its canyons are as deep as 4.5 miles, or 23,000 feet, which is four times deeper and steeper than the Grand Canyon of Arizona. In places the grabens are 120 miles wide, and in places it is 300 miles between them.

Here again, a descriptive phrase from A. B. Guthrie is recalled. These canyons and their near vertical canyon walls, in the "Serene Hemisphere" of Mars, are truly "high, wide and handsome."

> Visible even on long-distance images of Mars is the great canyon system, which straddles the globe just south of the equator between longitudes 30° and 110° W. Called Valles Marineris, this 4,000 km long network begins on the east side of the Tharsis Bulge and ends in an immense region of chaotic terrain between Chryse Planitia and Margaritifer Sinus. At its deepest it is some 7 km deep and individual canyons are up to 200 km. in width. In the impressive central section, where there are three roughly parallel, interconnecting rifts, the total width is 700 km. [n1]

The various grabens on Mars are parallel, even though 100, 200 and 300 miles apart. In Africa, similar parallel rifts - grabens and horsts - occur with spread of 300 miles in Kenya, Tanzania and Uganda. But Africa's rift valleys are only a half mile to a mile deep rather than four or five miles deep, the case here. In both cases, the systems are over 2,500 to 3,000 miles long, but for Mars that length for a rift system is 19% of the planet's circumference. The length of the African Rift Valley is only 11% of the Earth's circumference.

The gradualist menu, always of interest, on the genesis of the rift system of Mars is as follows:

Valles Marineris. We now know it to be a large fracture running out from the Tharsis Shield. This photo mosaic taken by the Viking Orbiter 1 very clearly shows the three massive shield volcanoes that sit astride the Tharsis Ridge - Arsia Mons, Pavonis Mons and Ascraeus Mons -- as well as the vast canyon system that is Valles Marineris. Similar features occur on Earth [East African Rift Valley], the Moon and Venus. Valles Marineris MAY HAVE BEEN FORMED AS MAGMA WAS WITHDRAWN FROM BENEATH IT TO SUPPLY THE VOLCANIC UPLIFT REGION. [n2] [Caps ours]

Something of the length, breadth and depth of this Rift System, the deepest in our Solar System, is ALSO in the following caption:

Scale of Valles Marineris. A comparison of, for example, the Tithonius Lacus region of Valles Marineris and the Grand Canyon shows the former to be four times deeper, six times wider and ten times longer. [n3]

Were the Valles Marineris in North America, it would stretch from Seattle to Central Florida. On another axis, it would stretch from San Diego to the capitol city of Quebec - or to Portland Me. or Boston. And its depth would be six times deeper than the Grand Canyon. By comparison, Valles Marineris makes both the East African Rift Valley and the Grand Canyon look tame.

The Genesis Of The Valles Marineris

In footnote 2 above, Moore and Hunt tentatively suggest "Valles Marineris MAY HAVE BEEN FORMED as magma was withdrawn from beneath it to supply the volcanic uplift region." Their suggestion is poorly conceived for the following four reasons.

1. No reason is proposed why magma should evacuate from ONLY under the Valles Marineris, but not from elsewhere.
2. If magma were to have evacuated slowly under the Valles Marineris, the crust would have wrinkled, not torn. When a person loses weight, their skin tends to wrinkle, not tear. The same applies to the crust of Mars. A slow withdrawal produces effects that would be quite different than what is evident on the red planet's crust.

3. In the next chapter, an analysis is made of the total mass contained in the top twelve volcanoes of Mars. It is from surrounding surface level to crater rim. Their total or composite volume will be shown to be in the range of 1,000,000 cubic miles.

The volume of the magma of Mars is over 40,000,000,000 cubic miles. The amount of volcanism evident on the Martian surface is only 0.0025% of the total magma of Mars. It is one part in 40,000.

4. The Valles Marineris gives the appearance of youthfulness, a relative lack of wind erosion.

One million cubic miles of magma, or lava, may be a huge amount by the standards of volcanism on the Earth. But by Martian standards, it is one six hundredth of the amount of mass that Mars gained from the fragmentation of Astra. It is not much by the standards of how much mass Astra's fragmentation added to Mars.

Mars experienced internal distress from the fragments of Astra. After that shocking event, Mars experienced a long series of perhaps 300 flybys of the Earth and Venus, all within 100,000 miles. Thus Mars experienced a blast. Next it experienced a long series of squeezes.

The volcanoes of Mars were a result of those sudden, speedy, swift squeeze plays, managed by the Earth and Venus. Tides within Mars, sudden massive magma tides are what caused its volcanism. Those tides were generated by the totality of the magma of Mars, not just from under one locality, and a locality 1,000 miles distant at that. The volcanism of Mars is simple to explain. The rifting of the crust of Mars is also simple to explain; the reasons for rifting are not the reasons for Martian volcanism.

The Great African Rift Valley And Sodom-Gomorrah

The Great Rift Valley of Africa appears to have had its genesis after the time of the last major shift of the spin axis and shift in location of the equator and its bulge zone. In Volume 3, a foundation will be laid that this event occurred after the closest of the many flybys of Mars. It was in the middle of the third millennium B.C.E. Sumerians called it "Utnapishtim's Flood, and reported that it was caused by Enlil (the Sumerian word for Mars). [n4]

It was only 500 years later, very early in the 2nd millennium B.C.E. Abraham was on the scene which involved the northern extremity of the GREAT AFRICAN RIFT VALLEY. It was still active, still splitting, still deepening, still widening, still erupting on its edges. Isostatic shifts of magma create a new equatorial bulge zone but they tend to be slow changes.

Abraham was positioned on the edge of the ledge, and above the rift valley which he could view below in the distance. It was called The Vale of Siddim, and included five cities below, the mouth of the Jordan River and the Dead Sea. It was a Mars flyby of the March case. It was over 500 years after Noah's Flood, and almost 500 years before Moses. The description in Genesis is as follows.

> The sun was risen upon the earth when Lot entered into Zoar.
> Then the Lord rained upon Sodom and upon Gomorrah brimstone and fire from the Lord out of heave;
> And he overthrew those cities, and all the plain, and all the inhabitants of the cities, and that which grew upon the ground.
>
> Genesis 19:23-25

Talmudic information is as follows:

> Lekah Gen., loc. cit., states that the angels came suddenly like lightning upon Lot, whereas Abraham discerned their arrival from afar. [n5]

It was a catastrophic scene drawn from the dawn of Hebrew history. It was an eye witness description; Abraham was there, viewing the Rift Valley's northernmost finger. It was the trench containing the Jordan River and the Dead Sea. In the distance were fertile irrigated fields below, and five prospering cities. Two of them were Sodom and Gomorrah. A prosperous society there was reported to have given itself over to dissolution and to unnatural sex orgies. All five cities were suddenly engulfed in volcanic gases, ash and a few lava flows. Those lava flows and cinders still remain.

The "death angel" was the triggering agent for this extension to the Great Rift Valley system. Simply put, it was another of the close flybys of Mars. In this case, the "death angel" had a mass 11% of Earth's mass. It had a cometary tail as shall be demonstrated in Chapter 7. It had two visible satellites, the sons of Arcturus in Hebrew, Deimos and Phobos in Greek. This angel had a name in Talmudic Hebrew texts; it was known both as the destructive angel of the Lord and as the evil archangel "Sammael." [n6]

The ancient Hebrew explanation for this event was the general wickedness and dissolute behavior in the five cities in the fertile plain at the bottom. Apart from the Hebrew explanation, and adding thereto, is that pressures had been building up for centuries, waiting for a trigger, to produce another extension in the Great African Rift Valley system.

A Mars flyby triggered the event. At that particular occasion, our estimate of the date was October, 1728 B.C.E. and the closest physical distance between the Great African Rift Valley and the Valles Marineris was, for a minute or two, some 35,000 to 40,000 miles.

The Book of Job is among the oldest treasures of mankind. Job's understanding of cosmology and celestial activity was a big issue in his time, and the issue was recorded as follows:

Figure 5 - The Valles Marineris on Mars

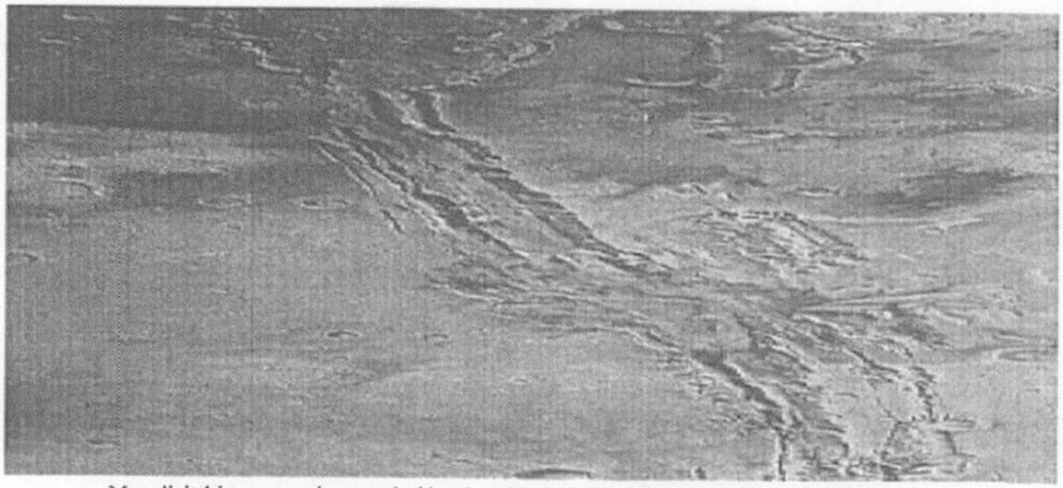

Mars digital-image mosaic merged with color of the MC- 18 quadrangle, Coprates region of Mars. Moderately cratered and faulted highland ridged plains in the northern and southern parts are cut by the prominent Valles Marineris chasma system, which reaches depths of 10 km and extends in an east-southeast direction for about 2,500 km across the quadrangle. The long, central canyons appear to be large, fault-bounded rifts, whereas some of the isolated, northern canyons are the sources of large outflow channels. Latitude range -30 to 0 degrees, longitude range 45 to 90 degrees.

Figure 6 - A Comparison of Dimensions - Grand Canyon verses Valles Marineris

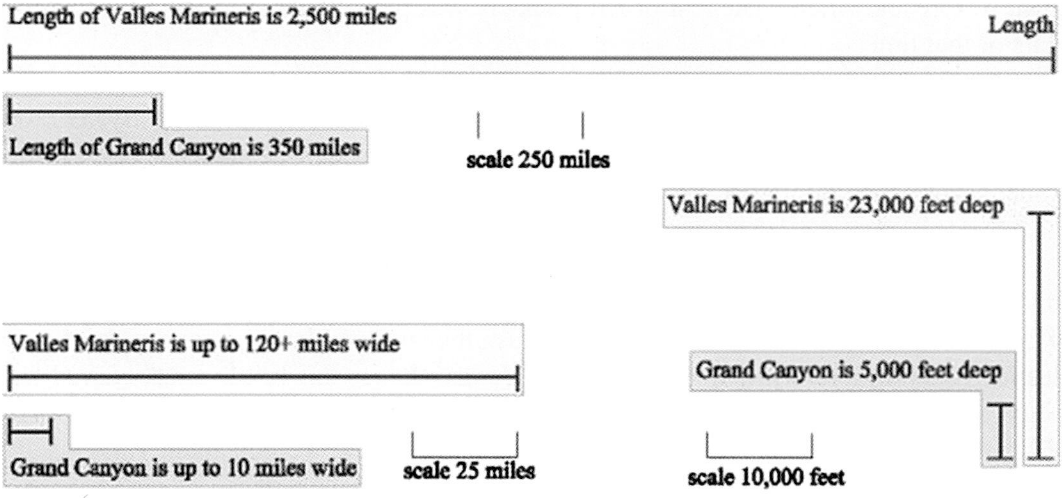

A Comparison of Dimensions - Grand Canyon verses Valles Marineris

Then Job answered and said, [God]

> Which removeth the mountains, and they know not, which overturneth them in his anger.
> Which shaketh the earth out of her place, and the pillars thereof tremble.
> Which commandeth the sun, and it riseth not: and sealeth up the stars
> Which alone spreadeth out the heavens, and treadeth upon the waves of the sea.
> Which maketh Arcturus [Mars], Orion [Jupiter] and Pleiades [Saturn] and the chambers of the south [Milky Way].
> Which doeth great things past finding out; yea, and wonders without number.
>
> Job 19:1, 5-10

Conclusion

This catastrophic cosmology addresses the increase in mass that Mars experienced as a result of absorbing some 30% of the fragments of Astra. This analysis is an increase in mass for Mars of about 1.5%, due to the blizzard of fragments hitting the surface of the Clobbered Hemisphere. As a result, in time, the physical diameter of Mars needed to expand by some 20 miles in the polar diameter and by 21 miles in the equatorial diameter.

Story 7 describes one of the results of Astra's fragmentation near Mars - MARS PUT ON A LITTLE WEIGHT. It is a story of a crustal expansion on a relatively small planet, and a nearby one. It is a story of a suddenly increase in mass, difficult to encase by the old crust. It tore and rifted. The tear or rift probably was perpendicular to the old equator of Mars of that time.

It is a story of the creation of the most gigantic rift valley in the Solar System, one replete with suddenly uplifted blocks, the horsts, interlaced with adjacent non-uplifted grabens, flat valley floors.

Those deep, steep, sharp canyons developed in response to pressures emanating suddenly from under the Martian crust following the blizzard of asteroids. With its radius of curvature and its estimated 20-mile thick crust, the crust of Mars was (and is) brittle, not elastic. It is far more brittle than is the Earth's crust. It was this brittle crust, curved with a high angle of curvature, that suddenly had to expand. The gigantic rift valley across the Martian crust was nature's response to Mars putting on a little weight, about 1.5%.

Probably these splits and uplifts occurred in a long series, beginning within weeks of Astra' demise. Extensions of the rifting occurred for centuries. Subsequent Mars-Earth flybys and Mars-Venus flybys were triggers for surges exerting more (momentarily expanding) sub crustal pressure.

The thick, brittle crust of Mars at first resisted those pressures to expand. But when its crustal elastic threshold did yield, and when its crust did begin rifting, it was major league rifting. The Valles Marineris accommodated an estimated 78% to 80% of the increase in the mass of Mars. Tharsis and Elysium, the two bulges, accommodated 20%. Volcanism, including all later celestial squeeze plays, is estimated at a mere 2%.

By comparison, on the Earth only minor league rifts have occurred. This is due in part to our crust's greater elasticity, and it is due in part to our planet having been buzzed by a smaller planet (not a larger one). Also, the Earth did not experience sudden increases in mass of one or two percent.

The Tharsis and Elysium bulges on Mars, like the Valles Marineris, were created by a sudden increase in mass. The Valles Marineris was due to long term pressures from a sudden increase in mass. Its appearance was delayed; it did not appear within hours like the bulges. The rifting process was slower, but if slower, nevertheless the rifting was inexorable. Rifting was the major adjustment made by the crust of Mars; bulging was the minor adjustment.

This 20th century AD is saturated with gradualists. Neither Abraham (20th century B.C.E.) nor Job (17th century B.C.E.), were gradualists. Nor was Moses. They saw what they saw. They were catastrophists. And if the dreaded angel of the Lord had a mass of 11% of the Earth, they were planetary catastrophists.

With story 7, the reader now is 34% of the way to the penthouse of cosmology.

CHAPTER 4

The Biggest Volcanoes In The Solar System

Intuition often runs far ahead of proof. One feels privately convinced of the truth of theorems of which one cannot offer a public proof.

E. C. Titchmarsh, speaking of mathematical proofs.

Introduction

This is the fourth of six chapters with attention directed to the physical geography of Mars and its two satellites - the scars from Astra's fragmentation - the evidence. This topic encompasses chapters 1 through 6.

Nothing would set off a cycle of magnetic surges, and an orchestra of volcanic activity more quickly than would a close planetary flyby. A sudden, invading gravity such as the Earth would disturb and disrupt the molecules of the magma within Mars. How much would the magma be disturbed and relocated? For rotating planets, tidal forces respond according to the inverse cube law. Tides vary according to the inverse of the distance cubed.

For planets like Mars and the Earth, this means that as the distance between the two is halved, the tidal surges therein, created by each in the other planet, are increased eight fold.

As distances were halved twice, tidal volumes proportionately increase 64-fold. As distances were halved three times, as from 240,000 miles to 120,000 to 60,000 to 30,000, tidal volumes respond by increasing (8 x 8 x 8) 512-fold. From 240,000 miles, the distance of the Moon to 15,000 miles, tidal forces would have increased 4,096 fold (512x8). These conditions would produce massive magnetic surges.

The Location Of The Volcanoes Of Mars

The momentary subject is magnetic tides within Mars, not within the Earth. Mars has an estimated crustal thickness of about 20 miles, twice the thickness of the Earth. Also its radius of curvature is greater, making its crust more rigid.

The crust of Mars has very little flex, or elasticity in contrast to the Earth's crust. Therefore, for Mars, with no crustal elasticity, its primary mechanism was to relieve internal distress through volcanism. This, plus the probability that Mars suffered several hundred Earth and Venus flybys, is why the volcanoes of Mars are so gigantic.

By the consensus of geologists, there have been 170 paleomagnetic polarity reversals recorded in Earth lava flows. A foundation will be laid that each flyby produced a geomagnetic field polarity reversal. Therefore there is some evidence that Mars-Earth flybys repeated, and recurred perhaps 150 to 200 times. There is reason to assume that Mars-Venus flybys were similarly numerous. Therefore there might have been 350 planetary flybys involving Mars all together.

On Mars, its gigantic volcanoes are all located in its Serene Hemisphere, just like the two bulges, the Tharsis Bulge, the Elysium Bulge and gigantic rift system, the Valles Marineris. These are all indications that the Martian crust is somewhat thinner in its Serene Hemisphere. On Mars, 90% of the volcanoes are located on the huge Tharsis Bulge, and 10% are located on the Elysium Bulge.

Basaltic traps occur when there are one or two vast bleedings of lava on a planet's crust. Basaltic traps are generally flat, and like lakes, sometimes are very broad.

On the other hand, volcanic cones occur when there have been many dozens, perhaps hundreds of successive lava outflows out of one caldera, or volcanic crater. They build up by repeated eruptions. Volcanic cones typically have gradients of 5%, 6% or even 7%.

In the Serene Hemisphere of Mars, gigantic volcano cones occur whereas very little ones are found on the Earth. The size of the Martian volcanoes is a reflection of how many and how close its ancient flybys were of the Earth and Venus.

The Sizes Of The Volcanoes On Mars

Sizes Of The Calderas (Ejecta Craters)

In the Serene Hemisphere of Mars are found the loftiest, broadest dozen volcanoes in this Solar System. They are few in number, and that is a reflection of how thick the crust of Mars is. But their gigantic sizes are a reflection of how many spasms of catastrophism the red planet experienced.

Vast volcanism is a reflection on which planet experienced the most internal distress, and which planet suffered the greatest number of crustal/magma squeezes. During those repeated squeezes, magma gushed out of its vents or calderas in vigorous volumes of white hot gushing lava.

BASALTIC TRAPS. The Columbia Plateau of Washington-Oregon-Idaho is such a basaltic plateau, covering 150,000 sq. miles of surface, including minor parts of Southern British Columbia and Northern Utah in addition. The vents were many and the bleeding was simultaneous. The basaltic Deccan Plateau of India is an even bigger example, and its genesis will be discussed in Volume III. Vents such as this are found in Arizona also. A basaltic trap, formed by one vast bleeding of magma, is not to be confused with a volcanic cone.

VOLCANIC CONES. In contrast, single, individual volcanoes are built up slowly and regularly by numerous layers from repeated eruptions of lava and ejecta from the same caldera. Volcanoes require multiple dozens, if not hundreds of eruptions and outflows of magma, to gradually build up their cones.

Two or three eruptions do not produce sufficient ejecta for a volcanic cone. Two or three dozen eruptions might create a small volcanic cone. A hundred eruptions from the same crater will build a volcano on the Earth; it was the same on Mars. As was mentioned earlier, volcanoes have steep gradients, with slopes having angles typically ranging around 6%. On the other hand, basaltic plateaus are flat, like lakes.

Volcanoes have deep vents, where lava outflows, or flows out at temperatures ranging between 2,000 and 3,000° F. In time, the magma cooled, hardened and became black. Volcanoes build up slowly, by repeated spasms, and by repeated up thrust flows through unplugged vents. When a vent becomes plugged with cold, solidified lava, it is called a "pipe." Sometimes volcanic pipes contain a sprinkling of diamonds.

CALDERAS (BLOW-HOLES). On the Earth, calderas, craters in the cone of a volcano, are usually but a fraction of a mile or two in diameter, though some have diameters up to three miles. Crater diameters are one indication of how much ejecta was expelled through its vent (or vents). Nothing on the Earth's surface compares to the caldera size of Martian volcanoes.

The caldera of Olympus Mons is measured at 50 miles in diameter. The caldera diameter another huge giant, Arsia Mons, is even wider at 65 miles. The caldera diameter Pavonis Mons is estimated at 20 miles. That of Ascraeus Mons is estimated at 24 miles. The caldera of Ulysses Patera is about 20 miles in diameter as is the caldera of Biblis Patera. Alber Tholus has a crater diameter estimated at 15 miles. Tharsis Tholus is estimated at 12 miles wide. By Earth standards, all of these calderas are immense vents, indications of vast volumes of lava outflows.

One could stuff all of metropolitan Los Angeles into the caldera of Arsia Mons. New York City and most of its suburbs could be stuffed into the caldera of Olympus Mons. Chicago, including all of Lake County and Du Page County also could be stuffed into the caldera of Olympus Mons. Two Philadelphia's or ten San Francisco's could be stuffed into the caldera of Pavonis Mons.

By contrast, on the Earth, the Mauna Loa caldera on Hawaii averages only three miles in diameter, and this is the Earth's largest volcano, rising from 20,000 feet below the surface of the Pacific Ocean to 13,700 feet above sea level.

The caldera of the famed Kilimanjaro is 1.5 miles in diameter. This volcano is only 13,000 feet high, rising from a base on a 6,000-foot plateau to 19,324 feet. The caldera of the famed Ararat is one mile in diameter. Ararat rises only 12,000 feet above the highlands of Eastern Turkey.

Compared to the volcanoes of Mars, volcanoes such as Kilimanjaro, Ararat, Etna, Popocatapetl, Shasta and Rainier are just five widely scattered pimples on different continents. The vast volcanoes of Mars, on the other hand, are much larger and are clustered into only two regions.

The Estimated Elevations Of The Volcanoes Of Mars

ELEVATIONS ON THE EARTH. Huge volcanoes on the Earth typically rise 6,000 to 10,000 feet above the surrounding terrain. Examples are the fore mentioned Kilimanjaro and Ararat. Other well-known examples include Rainier, Popocatepetl, Etna, Shasta, Hood, Baker and Cotopaxi. Mauna Loa-Mauna Kea on Hawaii, rising some 30,000 feet from an oceanic floor, is not typical.

ON MARS. On Mars, the highest volcano is Olympus Mons. Its elevation is estimated at 82,500 feet, or 15.6 miles. The height of Olympus Mons is measured from the surrounding plain; there is no such thing as mean sea level on Mars.

Ascraeus Mons is the second highest volcano in our Solar System, and has a crater rim about 50,000 feet above the surrounding plain. There is 5% to 7% grades for the slopes of Martian volcanoes. Arsia Mons is estimated at 40,000 feet, as is Pavonis Mons. They are third and fourth highest and widest in the Solar System.

The Volumes Of The Vast Volcanoes Of Mars

CONIC BASES. Olympus Mons has a conic base with a diameter of 325 miles, and an area of 101,000 sq. miles. Its volcano base area compares to the area of such states as Arizona, or Colorado or New Mexico. On the basis of conic volumes, Olympus Mons contains 525,000 cubic miles of ejecta materials, largely lava.

Ascraeus Mons has a volcanic base of 50,000 miles, more area than the state of Ohio. Its area also compares well with the areas of Pennsylvania or Bulgaria. Ascraeus Mons, at 50,000 ft., has a volume of nearly 150,000 cu. miles.

The cone of Arsia Mons has an area of some 29,000 sq. miles, and is 40,000 ft. high. Its volcanic base area is about equal to that of South Carolina. Arsia's volume is estimated at 73,000 cu. miles.

Pavonis Mons has a volcanic cone base area of 38,000 sq. miles, much like the areas of Hungary or Indiana. At 40,000 ft. high, its volume is 76,000 cu. mi.

The volume of these four giants totals some 825,000 cu. mi. If this volume of basalt were used to pave the state of Illinois, from Chicago and the Wisconsin line down to Cairo, the entire state of Illinois would be paved 14.5 miles high.

Figure 7 - Olympus Mons

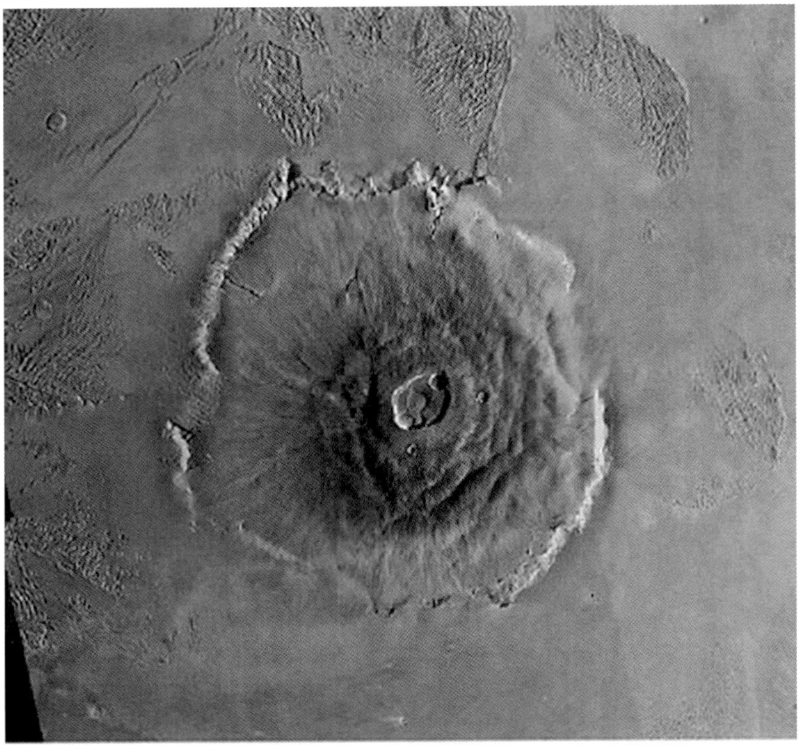

Shown here is a digital mosaic of Olympus Mons, the largest known volcano in the Solar System. It is 27 km high, over 600 km at the base, and is surrounded by a well-defined scarp that is up to 6 km high. Lava flows drape over the scarp in places. The summit caldera (central depression) is almost 3 km deep and 25 km across. Much of the plains surrounding the volcano are covered by the ridged and grooved 'aureole' of Olympus Mons.

Olympus Mons

Figure 7 is a photo of Olympus Mons, in the "Serene" Hemisphere of Mars. It is the serene side of Mars only for craters, but not for rifting or for volcanism. For comparisons in both height and breadth, Figure 8 compares the sizes of Olympus Mons on Mars, Mt. Kilimanjaro in Tanzania, and Mt. Ararat in Turkey.

Figure 8 - A Comparison of Olympus Mons to Volcanic Mts. Ararat and Kilimanjaro

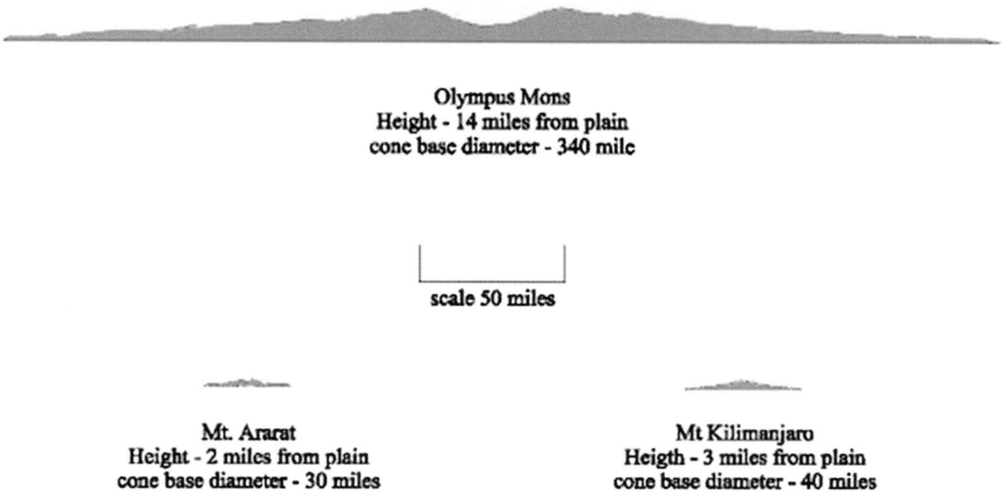

A Comparison of Olympus Mons to Volcanic Mts. Ararat and Kilimanjaro

By contrast the following table shows the volumes for some of the best known volcanoes on the Earth:

These four largest volcanoes on Mars spewed out gases, rocks and lava with a total volume of well over 800,000 cu. mi. This volume of basalt would pave all 48 continental states evenly to a depth of 1,350 feet.

Eysium Mons, in the other bulge on Mars, covers as much area as Switzerland or Belgium. Added to this total are the lesser lava outflows from Alba Patera, Albor Tholus, Biblio Patera, Hecates Tholus, Tharsis Tholus, Ulysses Patera, Uranius Patera, and Uranius Tholus. Thus the total volume of lava flows on Mars exceeds 900,000 cubic miles from its various volcanoes. The surfaces of the Earth and Venus have nothing to compare to this scope of volcanism.

Table V - The Volumes for Some of The Best Known Volcanoes on The Earth

Mauna Loa/Mauna Kea	4,700 cu. Mi.
Kilimanjaro	3,600
Cotopaxi	700
Popocatapetl	500
Etna	500
Ararat & Little Ararat	440
Rainier	200

The Viewing Of The Eruptions On Mars During Flybys

Ancient Greeks reported sighting and also, perhaps, the timing the orbits of tiny Deimos and Phobos. The three diameters of the fragment, Phobos, average 14.7 miles. Three of the diameters of even littler Deimos average 7.90 miles. Tiny or otherwise, nevertheless, the ancient Greeks did report seeing them, and apparently they (or more likely their Sumerian predecessors) named them. Whether or not someone in ancient times observed and sketched their orbits is a topic for Chapter 11.

Those furious, fiery eruptions on Mars, glowing in the dark, were perhaps a mile or two wide and were scores of miles long; some were over 100 miles long. The flowing rivers of lava on Mars were longer if not wider than Deimos and Phobos. And they started out white hot, and changed to yellow, orange, red and reddish black. Deimos and Phobos are low in albedo (reflectivity). Those rivers of lava probably glowed brighter than did the two satellites of Mars.

The caldera of Olympus Mons is 50 miles in diameter. The diameters of Deimos and Phobos are 10 and 15 miles in diameter. The caldera of Olympus Mons is almost 2,000 sq. miles in area. This is three times the diameter of Phobos and five times the diameter of Deimos. As the rivers of lava were longer than Deimos and Phobos, the calderas of the big four on Mars were wider also. If Deimos and Phobos could be seen by Greek eyes, so could they.

On Olympus Mons, those glowing rivers of lava began at the caldera, the crater. Rivers of hot, incandescent lava flowed for up to 180 miles apparently in all directions down its sides and onto the plain beyond. If they were seen by ancient Greek eyes, were they also reported?

Hesiod On The Appearance Of Ares In 701 B.C.E.

Hesiod saw the final flyby in 701 B.C.E. from Greece. He saw it in its celestial splendor, and in its frightful context also. He penned the following:

> There grew a hundred snake heads, those of a dreaded dragon,
> and the heads licked with dark tongues, and from the eyes on
> the inhuman heads fire glittered from under the eyelids;
> from all his heads fire flared from his eyes' glancing; [n1]
>
> On manslaughtering Ares, as he came onward, keeping his dread eyes upon him,
> Like a lion that has come on a victim, and with his strong claws, violently tears up the hide. [n2]

In Theogony, another work of Hesiod, he endeavored to put down the cosmic history of Mars scenes for perhaps 1,500 years. The following passage is about Thyphoeus (Typhoon), an archetype of Mars. This is the origin of our word "typhoon."

> The hands and arms of him are mighty, and have work in them,
> and the feet of the powerful god were tireless, and up from his shoulders,
> there grew a hundred snake heads, those of a dreaded dragon,
> and the heads licked with dark tongues, and from the eyes on
> the inhuman heads fire glittered from under the eyelids;
> from all his heads fire flared from his eyes' glancing;
> and inside each one of these horrible heads there were voices
> that threw out every sort of horrible sound, ... [n3]

At least seven times in Theogony, Hesiod referred back to the earliest times in Greek collective memory, which were the times of one "Iapetus". Iapetus has long been identified as the Hebrew Japheth of Genesis 8, a grandson of Noah. One of Japheth's sons was Javan, from which is derived "Ionian." And one of Javan's sons was Elishah, or Hellas. Greeks prefer to call themselves "Hellenes", after Hellas, a great grandson of Noah. "Hellas" is also the name of the largest asteroid crater in the Solar System.

Other families related to the Hellenes include Tarshish (Trojans), Kittim (Cypriots) and Dodanim (Dodecanese). Clearly, Hesiod assessed a historical era for the Greeks that was parallel to that of the Hebrews as occurs in early Genesis.

Job On The Celestial Scenery During A Mars Flyby

Job lived during a serious Mars flyby time, he also probably saw the rivers of lava flow from the calderas of Olympus Mons, Arsia, Ascraeus and/or Pavonis Mons during the devastating flyby of his era, so damaging to his land of Uz. The Book of Job describes an October case Mars flyby. Mars was the celestial dragon of the cosmos, going by the nickname of Leviathan, the serpent of the cosmos. It was during the October Mars flyby of the 18th century B.C.E.

> They are joined one to another, they stick together, that they cannot be sundered.
> By his sneezings a light doth shine, and his eyes are like the eyelids of the morning.
> Out of his mouth go burning lamps, and sparks of fire leap out.
> Out of his nostrils goeth smoke, as out of a seething pot or cauldron.
> His breath kindleth coals, and a flame goeth out of his mouth. Job 41:17-21

Apparently Job saw the volcanoes of Mars "awaken, like eyelids in the morning. When viewed from the Earth, and through the Earth's ashy, hazy, smoky atmosphere, those flows of hot lava may have taken on a reddish hue, due to the vast volume of contaminants beclouding our planets' atmosphere. People who have fought forest fires have observed at noontime a blood-red Sun and/or Moon.

Average Volcanic Flow Volumes Olympus Mons

Out of the caldera of Olympus Mons, over its active life, bled an estimated 525,000 cubic miles of lava and ejecta. Geologists estimate 170 paleomagnetic polarity reversals - their estimate may or may not be correct.

Geologists, gradualists in principle, despite all evidence to the contrary, do not realize that it was Mars flybys that caused paleomagnetic polarity reversals.

Even more important, they have failed to realize the collective Mars flybys were the ancient generator (dynamo) of the Earth's geomagnetic field, a field now dead. As a result of no more flybys, the Earth's geomagnetic field is dead with a residual decay rate of a half life of 1,350 years. A foundation for this knowledge will be laid in Volume IV.

Perhaps there were 350 Mars flybys of the Earth, some very close, some not as close to the Earth and Venus. A "close" flyby is considered to have been under 75,000 miles, planet center to center. To be discrete, 20% of the ancient Mars flybys were "mega-catastrophes", but the majority of 80% were in the 35,000 to 65,000 mile range. The position and resulting influence of Saturn, for instance, could readjust how close Mars came from flyby to flybys. A foundation for this conclusion will be laid in Volume III, entitled The Flood of Noah.

This estimate of flybys indicates that there were 350. Divide 525,000 cubic miles of lava by 350. Thus, during an "average" flyby of Mars, of either Venus or the Earth, 1,500 CUBIC MILES OF LAVA flowed out of the caldera of Olympic Mons alone. Flows from Ascraeus Mons, Arsia Mons, Pavonis Mons, etc. were in addition. What Job saw was an average flyby; what Hesiod saw was a mega-catastrophe.

The streams of lava on Olympic Mons may have been gushing in a stream 150 miles long before they began to cool down and ceased to flow during those frigid Mars nights, normal temperatures below -150° F. Beginning as white hot, cooling while flowing, they glowed at colors successively from white to yellow, orange, red, dark red and black.

Consider what kind of a river of lava would be made by a 1,500 cubic mile flow, or even 15 cubic mile flow. Consider what an impressive visual a reproduction of a flyby scene would make. If Phobos and Deimos were seen by ancient Greeks and their predecessors, it is probable that those streams of glowing, reddish, orange lava also were seen. Seen also were the eruptions preceding the lava flows. They were seen flowing down the sides of Olympus Mons, Arsia, Ascraeus, Pavonis and other Martian volcanoes. Lava must have flowed down the sides of Olympus Mons at velocities of 20, 30 and even 35 mph.

What does the Book of Job mean when it describes Leviathan: "Out of his nostrils goeth smoke, as out of a seething pot or caldron. His breath kindleth coals, and a flame goeth out of his mouth. Job. 41:21. ... "out of his mouth go burning lamps, and sparks of fire leap out." Translators and Bible readers have puzzled for a couple of millennia.

The distance between Mars and the Earth for the Final Flyby that Hesiod reported was much closer than an the average flyby. In chapters 9 and 10, an estimate is made of the Final Flyby, 27,000 miles from the Earth's center, and 21,000 miles from closest surface to closest surface. In Chapters 9 and 10, a method is developed for estimating the distances for the final flybys by Mars of both Venus and of the Earth.

Conclusion

Volcanic outflows comprise a measure of how much internal distress Mars experienced from flybys of the Earth and Venus. Paleomagnetic polarity reversals and 108-year cyclicism suggest how intensely and often Mars was squeezed? The astronomical history of Mars has been one of repeated, massive cosmic squeezes.

Perhaps the best comparison is the repeated firing of blast furnaces, daily producing fiery iron ingots. The difference between our blast furnaces and pouring of fiery iron and these blast furnaces of Mars is that on Mars, the magma was squeezed out of huge vents into huge rivers flowing up to 150 miles long across its cold surface. At blast furnaces, liquid iron is poured into small heat resistant molds, a little at a time.

The vast, vicious, violent, voracious, volcanic flows from Mars (Olympus Mons, etc.) are testimony as to how energetic and also how numerous those planet skirmishes were. The dating of the literature by Abraham and Job are testimony as to how repeated they were. A close analysis of chronological dates and flyby scenes can yield a 108-year cyclicism. Testimony by Hesiod and Isaiah indicate how remote they were in time .... billions of minutes ago, 14.2 billions of them to be more or less precise. .

Earlier, a foundation was laid to understand that the crust of Mars is somewhat thicker than the Earth's crust, being colder and being made of lighter materials. It is more rigid, less elastic. During Mars flybys, the Earth's crust could absorb some of its internal distress merely by elastic flexing of its crust (earthquakes). But the primary mechanism of relief of internal distress was different for Mars; it was volcanism.

The sizes of the volcanoes on Mars indicate how dreadful was the internal distress which Mars suffered in the Catastrophic Age. Job, Hesiod and other ancients merely caught glimpses of Mar's volcanism as it was erupting.

Story 8 of the catastrophic skyscraper is that during the Mars-Earth Wars, SUDDEN, MASSIVE INTERNAL DISTRESS WITHIN MARS WAS RELIEVED LARGELY THROUGH VOLCANISM. Olympic Mons was the primary vent. Ascraeus Mons, Arsia Mons and Pavonis Mons also were contributing vents, as were a dozen lesser volcanoes. Once again to paraphrase Guthrie, the volcanoes of Mars are definitely high, wide and handsome. In comparison to them, Kilimanjaro and Popocatapetl, Etna and Ararat, are next to nothing.

There are numerous aspects to scenes painted by the ancients who saw the flybys and left records of their views. Words were used describing aspects of planetary catastrophism. For instance, there are the eyes of Ares, red and bloodshot, flowing and glowing in the dark, spewing out clouds of smoke and streams of fiery fluid. Those red "eyes," glowing in the dark, probably were the calderas of Olympus, Arsia, Ascraeus, Pavonis, etc.

A review of Greek cosmo-mythology provides the one-eyed Cyclops, the evil Medusa, the speedy Perseus, the ugly Gorgons, the devastating Typhon, the swift Pegasus, the bloody Chrysaor, etc. All were among the Greek archetypes of Mars. The Greeks had many nicknames for Ares for its various flybys.

The genesis of the massive volcanoes of Mars were planetary flybys of the Earth and Venus. These ordeals of Mars were repeating, and as shall be demonstrated in Volume III, were cyclic in 108-year cycles. It was the long series of squeeze plays put on Mars by its two neighbors, the Earth and Venus.

Story 9 is that, just like Deimos and Phobos, while these Martian volcanoes were near the Earth, erupting, THOSE ERUPTIONS WERE VISIBLE TO THE ANCIENTS. They were described, albeit in horrible, ugly terms, in the literatures of the ancients. Those scenes they described were not very pretty, but they were painted fairly accurately - often by eye witnesses.

With story 9, the reader is now 37% of the way to the penthouse of planetary catastrophism.

CHAPTER 5

The Place In Space Where Astra Fragmented

> *When I wrote my treatise about the system [the Solar System], I had an eye upon such principles as might work with considering men for the belief in a Deity, and nothing can rejoice me more than to find it useful for that purpose.*
>
> Isaac Newton, Philosophiae Naturalis Principia Mathematica.

Introduction

The issue of this chapter is, "Where was that place in space where Astra fragmented"? There are several ways to locate a place in space. If one is dealing with planets or asteroids, the plane of reference is the "ecliptic plane." This is the plane on which all of the planets revolve save Pluto.

Another method of locating a place in space is simply to say Astra's demise, and the genesis of the asteroids, occurred on the Roche Limit of Mars. But this does not indicate where Mars was on the occasion.

Another method of locating the event relative to other planets and the fixed stars is to use coordinates of celestial longitude and latitude. Arbitrarily, the Earth's autumnal node is at 0°. The vernal equinox, March 21, the first day of spring, is at 180°. But this indication does not indicate how far Astra was from the Sun on the occasion of its fragmentation.

The distance from the Sun where Astra fragmented can be approximated by averaging the distances of the perihelion's of the fifteen largest asteroids, and establishing that "average distance" from the Sun. If this average agrees with the ancient orbit of Mars, progress is achieved.

But what is meant by "average" for the fifteen largest asteroids and their perihelion's? Are they to be each given equal importance, like the delegation to the Senate from each state? Or are their perihelion distances to be weighted, like our House of Representatives? In this case it is better to ascertain their weighted average.

In astronomy, usually distances from the Sun are measured in either kilometers or "au." One au. is 93,000,000 miles, an astronomical unit. In this volume, miles are the unit of measure. Conversion to au. or kilometers is easy.

Average Asteroid Perihelion Distances From The Sun

The first theory is that Astra fragmented on the Roche Limit of Mars. The second theory is that ancient Mars had a different orbit, more eccentric and less round than its modern orbit. The third theory is that when Astra fragmented, it was approaching its closest distance to the Sun, its perihelion. This means Astra had a greater velocity than Mars. Astra overtook Mars, so close as to create its own demise.

If this thinking is sound, it may be possible to generally locate where the demise of Astra occurred. The orbits including their perihelion's of the larger, more massive asteroids are known. What is the average distance of their perihelion's? What mode is to be used? One mode disregards the weight or the mass of the asteroid. Small asteroids and large ones would be of equal significance. As was mentioned above, the better method is to weight the results according to the masses of the various asteroids.

The method used here favors recognizing that a large asteroid is a bigger factor than a tiny one. Table VI lists the perihelion's of the fifteen largest asteroids. The largest asteroid, Ceres, dominates. The mass of Ceres is 57% of the total mass of all fifteen of the largest asteroids.

The largest three asteroids are Ceres, Pallas and Vesta. Their collective mass is 79% of the total of the fifteen largest. The average perihelion distance of the trio, Ceres-Pallas-Vesta thus dominates even more the collective average of the largest fifteen. Table VI provides this information in miles, which is more common, but less scientific than kilometers or au. It relates better to the general readers of English.

Table VI - Perihelion Distances Of The Fifteen Largest Asteroids

ASTEROID	DIAMETER IN MILES	MASS ESTIMATE	EST. % OF TOTAL MASS	PERIHELION DIST. MILES
CERES	622	.242	57.6%	237,150,000
PALLAS	377	.050	12.8	196,230,000
VESTA	334	.037	08.8	199,950,000
Hygeia	279	.022	05.2	264,120,000
Euphrosyne	229	.010	02.4	227,850,000
Interamnia	217	.010	02.4	239,940,000
Davida	200	.008	01.9	247,380,000
Cybele	192	.007	01.6	279,930,000
Europa	179	.006	01.4	255,750,000
Patienta	171	.005	01.2	262,260,000
Eunomia	169	.005	01.2	199,950,000
Juno	155	.004	00.9	184,140,000
Psyche	155	.004	00.9	235,290,000
Doris	155	.004	00.9	272,490,000
Undina	155	.004	00.9	276,210,000
Total			100.1	

The average perihelion of the Ceres-Pallas-Vesta trio is 226,650,000 miles. However, the average perihelion of the fifteen largest asteroids calculates at 232,200,000 miles. The perihelion of Ceres alone is 237,150,000 miles. The average perihelion distance of the fifteen above suggest that the perihelion of Ceres, today, is slightly farther out than the place in space where Astra fragmented.

Astra as a planet ceased to exist while Mars absorbed tens of thousands of fragments of Astra, and gained a little weight, perhaps 1.5%. It also is probable that Mars gained a little angular momentum and a little energy. While the Solar system lost a small planet, it gained 5,000+ tiny asteroids instead.

The Probable Distance Of Astra's Fragmentation

This study points to a distance somewhere between 225,000,000 miles and 228,000,000 miles for the fragmentation of Astra. See Table XIII, p. 2, col. 3. It is estimated that the old aphelion of Mars was 228,805,000 miles from the Sun.

The suspicion is that Astra fragmented at a distance of 225,000,000 miles, (or 2.46 au.) Large fragments such as Hellas, Isidis, Argyre, etc. cannot be assessed although they too were part of Astra. Table XIII indicates that the aphelion of old Mars was 228,805,000 miles. If correct, Astra caught Mars just a few weeks after its old aphelion.

This is how far out Mars was from the Sun when the red planet and Astra had their fateful ruckus ... their explosive rendezvous ... their final furious fragmentation. Figure 9 illustrates the place and the geometry of the fateful approach of tiny Astra, as it was overtaking another relatively small planet. The diameter of Astra is estimated at 1,600 miles, somewhat larger than Pluto, but only 74% of the diameter of the Moon.

Aphelion Distances Of The Fifteen Largest Asteroids

THE PROBABLE APHELION OF ASTRA. Can the average of the fifteen largest asteroids also indicate where the aphelion of ancient Astra's orbit was? Probably. Ceres, the largest, has an aphelion of 377,400,000 miles. As was mentioned earlier, Ceres alone has more mass than the next largest fourteen combined.

When the average aphelion of the fifteen largest is calculated and weighted, it is likely that the aphelion of Astra was between 400,000,000 and 405,000,000 miles from the Sun. This is 82% to 84% of the way out toward Jupiter. In astronomical units, Astra's ancient aphelion is estimated at approximately 4.35 au.

Figure 9 illustrates the most probable geometry of approach for Astra and Mars, on that ancient collision course. Mars, at the slowest part of its ancient orbit, was beginning to return to the inner, warmer regions. Astra was on its way back out to a cold region 4.3 times the Earth's distance from the Sun. In this geometry, Astra's velocity was considerably higher than that of Mars. Thus it was Mars which was overtaken by Astra. There is a some evidence, and a probability that the place where Astra fragmented was in late Cancer - at about 290° celestial longitude. Thus the location of where Astra was converted into asteroids, is given (1) by celestial longitude, (2) by distance from the Sun and (3) by distance from where Mars was at that time. Astra was 2.5 radii from the center of the red planet.

Evidence For The Rotation Of Astra

Little Astra, like most other planets including Pluto, apparently had rotation. The spin rates for eight of the asteroids, in hours, are known. Their spin rates are as follows, in descending asteroid size:

Table VII - Spin rates for eight of the asteroids

Ceres	9.07 hours
Eunomia	6.08
Vesta	5.34
Juno	7.21
Davida	5.17
Psyche	4.30
Hebe	7.74
Nemausa	7.78

Astra's diameter is estimated at 1,600 miles plus or minus 100. When Astra rotated, those parts of it farther from its spin axis, in its equatorial zone, rotated fastest. Probably, the spin rate of Astra was about five hours. Materials close to the spin axis rotated slowly; materials in its equatorial crust rotated rather rapidly.

When Astra fragmented, the spin components of its various fragments was passed on to its fragments. Thus it follows that when the gigantic Hellas, Isidis and Argyre fragments hit the crust of Mars, and penetrated through its crust, they did so while still rotating. They were rotating at spin rates around 9.5 hours, much faster than Mars, at 24.6 hours.

These large fragments passed through the crust of Mars into the Martian innards - its magma with original spin still in tact. The Martian magma thus had to absorb not only the additional mass and the additional energy of these three fragments but their additional angular momentum (rotation) as well. Mars took three very hard shots that day, rotating Hellas, Isidis and Argyre.

Astra's diameter is estimated at 1,600 miles plus or minus 100. When Astra rotated, those parts of it farther from its spin axis, in its equatorial zone, rotated fastest. Probably, the spin rate of Astra was about five hours. Materials close to the spin axis rotated slowly; materials in its equatorial crust rotated rather rapidly.

When Astra fragmented, the spin components of its various fragments was passed on to its fragments. Thus it follows that when the gigantic Hellas, Isidis and Argyre fragments hit the crust of Mars, and penetrated through its crust, they did so while still rotating. They were rotating at spin rates around 9.5 hours, much faster than Mars, at 24.6 hours.

These large fragments passed through the crust of Mars into the Martian innards - its magma with original spin still in tact. The Martian magma thus had to absorb not only the additional mass and the additional energy of these three fragments but their additional angular momentum (rotation) as well. Mars took three very hard shots that day, rotating Hellas, Isidis and Argyre.

INTERNAL DISTRESS FOR MARS. When Astra fragmented, Mars suffered GREAT INTERNAL DISTRESS. The Tharsis and Elysium Bulges, and its Valles Marineris rift all reflect an initial, sudden internal distress within Mars. Spin rates for these fragments increased its internal distress, already at epic proportions. It is no wonder that the crust of Mars soon rifted in order to accommodate its new conditions.

On the other hand, the Martian volcanoes, Olympus Mons, Pavonis Mons, Arsia Mons, Ascraeus Mons, etc. reflect the subsequent series of internal distress of Mars during the Mars-Earth flybys and the Mars-Venus flybys. Volcanoes are built by many successive eruptions. Bulging and rifting were the primary products of absorbing 30% Astra's fragments. Volcanism was the primary method of relieving its internal distress during Mars-Earth flybys and Mars-Venus flybys.

Figure 9 - The Place in Space Where Astra Collided With Mars (And Mars Acquired its Catastrophic Third Orbit)

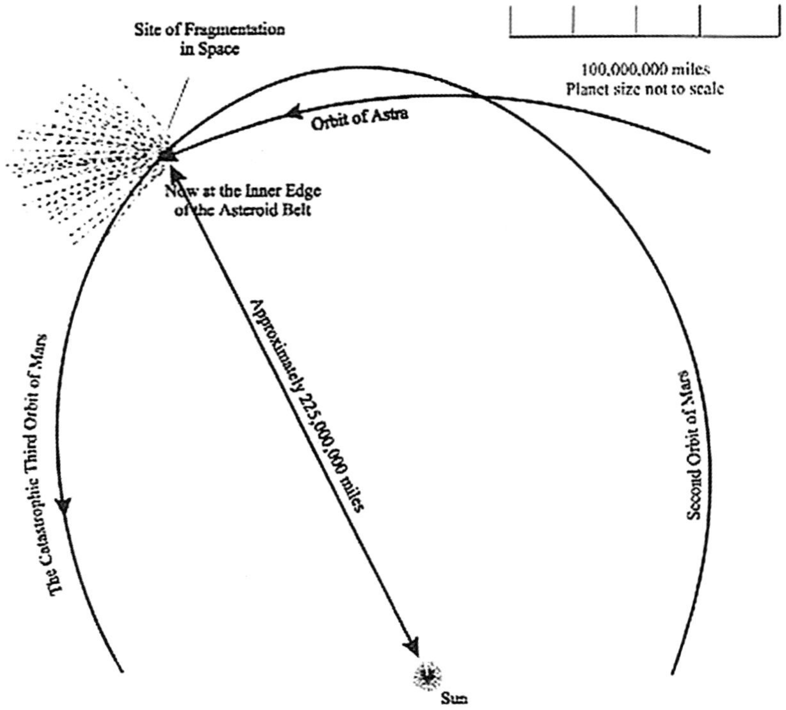

The Estimated Geometry of the Approach of Mars to Astra
Mars in a Collision Course Overtaking Astra
Site of the Sudden Birth of the Asteroids

Thus, there were three different methods of relieving internal distress for Mars, (a) bulging, (b) rifting and (c) volcanism. In addition, there were three different causes of its internal distress, (A) Astra, once (B) Venus, perhaps 100 times, and (C) the Earth, perhaps 150 times. Is it any wonder that the collective Scars of Mars make a good subject for science, for ancient history and for cosmology?

Dating The Fragmenting Of Astra

There is some evidence in Greek mythology as to the dating of the fragmentation of

Astra. But there is a better method to date that event, a method not yet used.

Your authors do not know whether Astra fragmented at the onset of the Mars-Earth Wars, or possibly it fragmented midway during these ancient cosmic skirmishes. More research on this subject is needed, and it should not be that expensive. The necessary equipment already exists in many places.

This study points to the probability that Astra fragmented about 225,000,000 miles from the Sun, at about 290° longitude, on the Roche Limit of Mars. While the date is not known, ancient Greek mythology may provide a clue as to when it happened. They have accounts that their early pre-flood ancestors saw the event occur in the heavens.

A Solution By Computer Simulation

There are some 5,000 asteroids of which the orbits for at least a thousand are known. In astronomy, it is possible to program orbits of asteroids 20,000 years or more into the future. Similarly, a computer simulation can be programmed to retro-calculate backward.

The theory is that if this model is correct, and the orbits of the 200 largest asteroids are retro-calculated, at some specific time and place, they will tend to congregate. This place will indicate (a) where Astra fragmented, and (b) when it fragmented.

The problems are:

a) Acquiring orbital data on the 200 largest asteroids

b) Programming the perturbations caused by Jupiter

c) Programming the perturbations caused by Saturn

d) Programming the perturbations caused by Uranus

e) Programming the perturbations caused by Neptune

f) Acquiring the appropriate computing equipment

g) Acquiring the expertise

h) Acquiring the funding and the time

PREDICTION I. This team has confidence in the theories and sub-theories of this research. It is expected that a retro-calculation of the orbits of the asteroids will reveal them congregating at some particular place and on a particular date in the past. That will be the place and the date that where and when Astra fragmented. It will be the date when Mars became a battered bastion of catastrophism.

PREDICTION II. The second prediction is that when this retro-calculation of the asteroids is completed, that date will be neither earlier than 15,000 B.C.E. nor later than 3,000 B.C.E. This is but a tiny 12,000-year window out of multiplied millions of years in the gradualist time scale.

PREDICTION III. The third prediction is that the place of the debacle of Astra in the cosmos will be within 5,000,000 miles, plus or minus, of 225,000,000 miles from the Sun.

PREDICTION IV. The fourth prediction is that the location of the fragmentation of Astra will be between 280° and 300° of celestial longitude.

PREDICTION V. Prior to 15,000 B.C.E., the Solar System had ten planets, not nine, from Mercury out to Pluto.

Time and skilled computer simulation will tell if these predictions are true. There is a 7% possibility that the predicted longitude zone will be correct "just by chance". There is another 10% chance that the predicted distance of Astra's fragmentation will be correct "just by chance". There is a 0.7% chance both will be correct by chance. More will be said about "chance" and gradualism in Chapters 11 and 12.

Merrily, the gauntlet is thrown down. Such a challenge should test the grit of gradualism. Dare the gradualists and the semi-gradualists accept the challenge to retro-calculate for the fragmentation of Astra? At risk is the paradigm of gradualism in the history of the Solar System.

The Fragmentation Of Astra In Greek Mythology

The ancestors of the Greeks remembered and kept the story of an ancient event in the heavens, an explosion of something. It reads like the fragmentation of Astra. Their "timing" reference is loose; it was sometime during their first age, their "golden age." This was their era before Noah's Flood when it was a different world, with different latitudes and climates.

In Volume III, a foundation will be laid dating Noah's Flood in one October afternoon in 2484 B.C.E. That was when the "golden age" ended, succeeded by the ages of silver, bronze and iron.

The golden age was a time when the climate of the Earth, or at least of North America, and Northern Eurasia was more temperate, when the vegetation was verdant and when men commonly would live long enough to enjoy their great-great grandchildren, according to some ancient sources including American Indian legends. A foundation for the "Golden Age" may be laid in a later volume.

It was during the golden age when the Sumerians mapped the heavens into 360°, and divided it into 30-degree divisions. The Greeks called it their "zodiac." "Zoa" was their name for the animals, of which seven or eight were nominated as signs of the various 30-degree zones. In their era, the fixed stars did not move; just about everything else might, including mean sea level, the crust and the cardinal directions.

The account of the fragmentation of Astra comes from Greek cosmo-mythology. It is suspected that there were earlier versions in Hittite, in Sanskrit and perhaps in Sumerian.

> **ASTRAEA.** Astraea, in ancient Greek mythology, the "star maiden," daughter of Zeus and Themis, or of Astraeus the Titan and Eos ("the dawn"), in which case she is identified with Dike. During the golden age she remained among men distributing blessings. But when the iron age began, she left the earth to disgust, and was placed among the stars as the constellation Virgo. She is represented with a pair of scales in a crown of stars. [n1]

Lempriere draws from many of the same sources in his description of Astra.

> **ASTRAEA,** a daughter of Astraeus, king of Arcadia, according to others, of Titan, Saturn's brother, by Aurora. Some make her daughter of Jupiter and Themis, and others consider her the same as Rhea, wife of Saturn. She was the goddess of Justice, and lived upon the earth, as the poets mentioned, during the golden age, which is often called the age of Astraea, but the wickedness and impiety of mankind drove her to heaven in the BRAZEN AND IRON AGES. [n2]

Thus it is that ancient Greek cosmo-mythology discusses something much like the fragmentation of Astra in their heavens. It is possible that there was a confusion between the Astra fragmentation and the Glacis fragmentation, discussed in Chapter 7.

This team suspects that Astra fragmented sometime between 15,000 B.C.E. and 3,000 B.C.E. Greek mythology favors a later date. Greek mythology may or may not be correct. Thus a carefully operated computer simulation program will involve programming asteroid positions and asteroid orbits for up to 20,000 years past, at the very least and possibly further.

One final perspective exists on which further commentary is needed. The ancient Greek accounts of the fragmentation of Astra have it that the celestial explosion and disappearing of a former star occurred in the zone of early Virgo, 330 to 360°. This model has the location of the fragmentation of Astra in Cancer, at 290 to 300° longitude. Greek mythology and the conclusion of the authors are 40° apart.

Ancient Hellenic Eyeballing Of Astra

She was, however, the last of the divinities who retired from the habitations of men, and after her return to heaven she was placed among the constellations of the zodiac, under the name of Virgo or Erigone. She is represented with a stern, but majestic countenance, holding a pair of scales in one hand, and a sword in the other. [n3]

If the estimate of Astra's diameter is 1,600 miles, its reflective surface was 38% of the diameter of Mars. When .38 is squared to compare Mars surface area and Astra disc area, it equals .15. If Astra was of similar reflectivity (albedo in astronomical parlance) as Mars, it was 15% as bright as Mars at equal distances. Could Astra have been seen by the naked eye at 225,000,000 miles? Yes, with difficulty.

At 225,000,000 miles, on a collision course with Mars, tiny Astra would have been difficult to see. The reported existence and sighting of ancient Astra is a two-edged topic. First, could the Greeks (who call themselves Hellenes) possibly have seen it so far away?

Second, why would the Greeks and their antecedents claim there once had been another planet in the cosmos unless they had seen it? Third, why would they have described the demise of a fictitious former planet in an scenario that was both sudden and explosive?

Astra's demise is a two-edged topic on yet another plane. It may be that the ancient Greeks were mistaken totally in their story of Astra. But then, it also may be that modern traditional 19th and 20th century astronomers have been mistaken on their guesses about the genesis of the asteroids. In this case, it would not be the first time the ancients, their records and their stories, have been underestimated by modern skeptics. Nor will it be the last.

Hebrew cosmo-mythology has several parallels.

> The pillars of heaven tremble and are astonished at his reproof.
> He deviated the sea with his power, and by his understanding he smitten through the proud
> By his spirit he hat garnished the heavens; his hand hat formed the crooked serpent.
> Job 26:11-13

The crooked serpent, or dragon was Mars with a cometary tail. See chapter 7. In

Chaldean cosmology, and in Hebrew cosmology, the crooked serpent went by a variety of names, Rehab, Asp, Teammate, Behemoth, Leviathan, etc. In Greek cosmology, Ares had more than fifty nicknames, Lotan being one. Lotan is cognate with Leviathan.

Conclusion

The average distance to the Sun of the perihelion's of the 15 largest three asteroids is about 232,000,000 miles. A bit closer in, at 225,000,000 miles, approximately, is a place in space where Astra fragmented. Those fragments which missed Mars continue to orbit, mostly in the region between Mars and Jupiter. Over 5,000 of them are known to exist.

Story 10 in our skyscraper of planetary catastrophism is THE PLACE IN SPACE THAT ASTRA FRAGMENTED. It was about 225,000,000 miles from the Sun, on the ecliptic plane, PLUS OR MINUS 5,000,000 MILES. Of the twelve zones of the zodiac, the evidence favors that the fragmentation occurred in late Cancer, at some 290 to 300° of longitude.

Ancient Greek cosmology relates the time Astra (Astraea) fragmented; it was seen by Greeks or their ancestors in the heavens, and disappeared reportedly in Virgo. Astraea is represented by a celestial tiara, a crown of stars and a pair of scales. [n4]

Her ancestry is confused. Some reported she was a niece of Saturn, the daughter of Titan and Aurora. Her place was in the heavens during the golden age, sometimes called the "Age of Astrae." She was the last of the (planetary?) deities to retire from the heavens.

Story 11 is that the aphelion distance of the ancient orbit of Astra can also be estimated from the aphelion's of the asteroids. ASTRA WENT OUT BETWEEN 400,000,000 AND 405,000,000 MILES FROM THE SUN. This is 84% of the distance from the Sun to Jupiter. By contrast, the modern orbit of Mars is only 30% of the distance to Jupiter.

The "ancient Greeks" consider themselves to be "Hellenes." They consider themselves to be descendants of Hellas, who was a grandson of Iapetos, identical with Japheth in the Genesis account in Hebrew. Thus, Hellas, patriarchal father of the Greeks, was a great grandson of Noah. Greek mythology goes back virtually to accounts passed on by passengers of the Ark.

With story 11, the reader is 40% of the way to the penthouse of planetary catastrophism.

CHAPTER 6

The Ancient Ring System Of Mars

> *The brighter the light that a new theory sheds on its proper field, the darker and more distorting the shadows in which it submerges the experience which lies around the borders of its true scope.*
>
> M. Polanyi

Introduction And Review

In Chapter 1, three categories of evidence were given indicating a fragmentation of a small planet on the Roche Limit of Mars. The first of these categories was the Hemisphere of Craters versus the Serene Hemisphere. A second category was the giant Hellas Crater, in the bulls eye zone of the Clobbered Hemisphere. It is the largest crater in the Solar System.

The third category was the Rim of the Hemisphere of Craters. By identifying the rim, one can locate the center of the Clobbered Hemisphere. Its center is 45° S. lat. and 320° W. long. From this central location, 93% of the craters of Mars are in that particular Hemisphere, and 7% are in the opposite hemisphere.

On Mars, one goes from one of the most densely cratered hemispheres in this Solar System directly to the opposite hemisphere, with 7% of the Martian craters. It, the Serene Hemisphere of Mars, centered at 45° N. lat. and 140° W. long., is easily the most serene, uncratered hemisphere in the Solar System.

In Chapter 2, one additional category of evidence of a fragmentation of a small Pluto-sized planet was presented. It was the two bulges in the Serene Hemisphere, with particular attention to the Tharsis Bulge. It is close to being directly opposite to the Hellas Crater.

The Hellas missile, perhaps 600-miles in diameter, hit the 20-mile thick crust of Mars at a velocity estimated at 25,000 mph. This also is 420 miles per minute, or 7 miles per second. The Hellas Fragment and at least two more, Isidis and Argyre, penetrated the crust of Mars and plowed into its hot, fluid magma. Sudden, immense pressures from these caused the two bulges to suddenly rise in the Serene Hemisphere. The bulges began to rise within 90 to 120 minutes after Astra fragmented.

In Chapter 3, it was determined that the acquisition and absorbing of these tens of thousands of fragments of Astra added something to the mass of Mars. Much of that increase was in the twenty largest fragments. That increase in mass for Mars is estimated at a sudden increase in mass for Mars of 1.5%.

Mars put on a little weight. Its other bulge zone, that is, its equatorial diameter, increased by 21 miles and its polar diameter increased by an estimated 20 miles. The crust responded appropriately as suddenly it was inadequate to contain the red planet's new mass.

Its crustal response was to rip and tear in a new, wide, long, 2,500-mile split, a rift, a torn crust, in order to accommodate a new, larger diameter. The rift system on little Mars is far wider, longer and much deeper than is the Great African Rift System in our Eastern Hemisphere.

As in the case of the Earth and its Great African Rift Valley, the Valles Marineris also probably formed at the perpendicular angle to the old equator of Mars. It was when Mars was in its catastrophic third orbit. Like the two bulges in the crust of Mars, the Valles Marineris is located in the "Serene" Hemisphere.

Like the Great Rift Valley of East Africa on our planet, the central region of the Valles Marineris also is the widest, deepest and most complex. An increase in mass caused the Valles Marineris on Mars. A shift in the equatorial bulge zone, and a relocated spin axis is what created the Great African Rift Valley.

The Tharsis Bulge and the Valles Marineris are the fourth and the fifth categories of evidence supporting the concept that the Roche Limit of Mars is what caused Astra to fragment. It is all rather straight forward. These five categories ALL ARE CLUES as to the cause of the tortured surface of Mars, and they point in harmony.

In Chapter 4, an additional sixth clue occurs. It is a testimony not so much as to the fragmentation of Astra, but to the ancient Catastrophic Orbit of Mars. Mars flybys of both the Earth and Venus occurred, apparently over 100 in each case.

On each close flyby of the Earth or Venus, the magma of Mars was again squeezed upward, and much lava and ash was vented from its giant volcano craters. The evidence is in the display of immense volcanoes on little Mars. These huge volcanoes illustrate the degree of internal distress Mars must have experienced during its ancient flybys of the Earth and Venus.

Also, the giant volcanoes of Mars support the idea that volcanism rather than crustal flex was the primary mechanism for relief of internal distress for Mars during planetary flybys. These are all "scars of Mars," all scars readily visible on its surface, its physical geography.

Chapter 5 addresses the seventh clue, those fragments which missed the crust of Mars, and proceeded on into space. Most asteroids went out toward, but not as far as Jupiter's orbit. They became asteroids when in their virgin orbits and they began to orbit the Sun. Those that have been seen number over 5,000. They also are scars in the Solar System of the fragmentation of Astra on the Roche Limit of Mars.

Most, perhaps 98% of the fragments that missed Mars began to orbit the Sun. It is theorized that the other 1% or 2% found Mars as their "centerpiece." They began to orbit Mars instead of the Sun, and formed an ancient ring system parallel with the Martian equator.

In contrast to the Sun asteroids, it is the Mars asteroids, comprising its former ring system, which provide our final battery of a half dozen more clues supporting the fragmentation of Astra on the Roche Limit of Mars. This, an ancient ring system of Mars, is the topic for Chapter 6.

How compelling are the first six categories of evidence? To that array of evidence is added what may be even more compelling, Deimos, Phobos and the former Mars asteroids. Gradualist astronomers do not recognize the evidence of a former ring system of Mars. But then again, they have yet to recognize, much less evaluate, the evidence that is now being placed in court.

A Blizzard Of Exploding Fragments And Logic

Astra fragmented into a spray of fragments. An estimated 30%, possibly 35% hit the surface of Mars in one hemisphere only. The rest of the fragments, some 65% more likely, 70%, missed Mars. They proceeded into 5,000+ new, virgin orbits.

There were three centers of gravity available for the virgin fragments as they needed to adopt a focus for their new orbit. The options for a new focus were the Sun, Mars itself and Jupiter. Our best guess is that between 98% and 99% began to orbit the Sun. 1% or 1.5% began to orbit Mars itself. And between .2% and .3% began to orbit Jupiter. Each fragment, indeed had to be snared by one of the three.

The vast majority of the fragments had sufficient energy, inherited from Astra's motion, to orbit out beyond the zone of control of Mars. But some fragments lost energy as the vector from the explosion opposed the orbit vector for the new fragment. Lacking sufficient energy to escape Mars, they began to orbit Mars. Such is the theory; the following is the evidence for this theory.

In astronomical parlance, this zone of control is known as the "radius of action." For the Earth, the radius of action is 750,000 miles, beyond which the Earth cannot retain a satellite. For Mars, this zone of control was much smaller, especially when Mars was near

Venus, and its ancient perihelion at a toasty 64,000,000 miles.

For all fragments that had their "aphelion" or "apoMars" within the Martian zone of control, they would become Mars asteroids. Fragmentation theory logically concludes that some, a small percent would stay, would circle Mars instead of the Sun. They are called "Mars asteroids," and thus differentiated from "Sun asteroids."

"Mars asteroids" is a new category of thought to astronomy because no astronomer of whom we are aware yet has considered, much less agrees that Astra fragmented on the Roche Limit of Mars. Nevertheless, it is the conclusion to which the evidence points.

The Evidence Of Mars Asteroids

Planets, since they rotate, are not quite spheres. Technically, they are "oblate spheroids." Due to spin rates and centrifugal force, they have an equatorial bulges and a pair of polar flat spots. Mars has a rotation rate only 41 minutes slower than the Earth's.

The equatorial diameter of Mars is 4,212 miles. Its polar diameter is only 4,183 miles, which is 29 miles shorter. The Earth's polar diameter is 27 miles shorter than its equatorial diameter. Mars has an oblateness of 0.009 compared to only 0.003 for the Earth. Thus Mars has an equatorial bulge that is a bit larger than the Earth's.

When satellites revolve around a planet, they will gradually align with its equatorial plane since that is where significantly more mass is located. Therefore, Saturn's rings are an extension of its equator, and so, if Mars had rings, the rings system of Mars also aligned in nearby space as an extension of its equator.

Deimos and Phobos do orbit on the equatorial plane of Mars today, as one would expect. Thus it was that the equatorial plane would have been the "zone of gathering" for any Mars asteroids including any lesser debris.

There is now a count of over 5,000 asteroids that have been found orbiting around the Sun. Almost all orbit between Mars and Jupiter. In contrast, Jupiter has at least twelve small, irregular-shaped satellites which appear to be former asteroids, which it captured on the fly. Jupiter's mass is almost one thousandth that of the Sun, and it is 2,958 times the mass of little Mars.

The first asteroid was discovered on the night of January 1, 1801 by Giuseppe Piazzi. It was first night of the 19th century. Many celebrated New Year's eve too much and too late and had hangovers the next night. But Piazzi planned his time better, preferring to scan the celestial regions alone with his new telescope. In the last 195 years, over 5,000 have been sighted, and dozens more are being discovered every year.

If 2% of the fragments of Astra orbited Mars, that still was a lot of debris. But if that much debris once was there, where did it go? Every time Mars made a close flyby of the Earth, some of the debris was swept out. The Earth swept them out like a broom. Every time Mars made a close flyby of Venus, more was swept out. Venus was a second broom.

If there was in excess of a hundred Mars-Earth Wars and another hundred plus Mars-Venus flybys, the former ring system of Mars could be depleted in due time --- swept nearly clean --- but not quite completely.

There exist a small number of tiny, irregular-shaped "inner asteroids," orbiting in the regions of the orbits of the Earth and Venus. Their origin seems to have been in the Catastrophic Era, and from the ring system of Mars. They seem to have been swept away from Mars. The names of these "inner asteroids" include Adonis, Amor, Apollo, Eros, Gaspra, Hermes, Icarus and Moore.

They are irregular, fragmentary in shape, just like the main belt asteroids. Their diameters are from a fraction of a mile up to five miles. They seem to be sweepings from the Mars-Earth and the Mars-Venus Wars. If they once orbited Mars, and were swept out into the inner Solar System, they should be reclassified as "Mars Asteroids".

Galileo was the first, at least in modern times, to see the sparkling ring system surrounding Saturn in 1610.F1 Among many of his discoveries with his new telescope, Galileo discovered the ring system of Saturn. It was close in to Saturn, at about Saturn 2.5 radii. This is the Roche Limit distance.

It is legend that for his discoveries, and his publishing of them, and talking about them, Galileo was summoned to the Vatican, the authorities of his time, and was tried by the Inquisition. He abjured the obvious and saved his life. Nevertheless others in Northern Europe also had telescopes, and they seconded his discoveries with enthusiastic reports, which were soon verified and reverified by others with their new telescopes. It was hard, even in the early 17th century, to keep good ideas down.

In the process of time, space mission cameras have discovered coal black rings, at the Roche Limit, 2.5 radii, from the center of Jupiter, circling Jove. Other coal black rings have been discovered circling Uranus and Neptune as well. These facts suggest that fragmentation's during near flybys, too near, have been repeated scenarios in time past.

In 1850 Edouard Roche, a Swiss physicist, determined that if two planets were on a collision course, assuming equal densities, the smallest of the two would fragment before a collision. Its internal tides would rise on a graph to the point of infinity, and fragment first. He found a small body would fragment at 2.44 radii from the larger of the two.

Roche assumed circular orbits and approaching planets of identical densities. Circular orbits occur seldom, and equal densities never occur in the real cosmos. Therefore the

"Roche Limit" is an approximation; it is valid but is subject to some alteration when density of the smaller approaching body is less than the larger planet.

There is no question but that Mars also could have had a ring system. The three questions are:

1) Did it have a ring system?

2) If so, where did that ring material go? and

3) Is some tiny debris still there, undetected?

It is likely that not all debris was swept out, and some still orbits near the orbit of Phobos, at 2.7 radii. Also, the debris would have to be small to remain undetected. This is expected to be a future discovery by space mission cameras near Mars.

Evidence # 7 - The Orbit Of Phobos

Tiny Phobos orbits 5,700 miles above the center of Mars, and it orbits on the plane of the Martian equatorial bulge. The radius of Mars is 2,110 miles. The Roche Limit of the red planet surrounds Mars at 5,150 miles from its center. Phobos orbits only 550 miles above the Martian Roche Limit. Phobos is fragment-shaped. Three of its diameters are 12.5 miles by 15 miles by 17.5 miles.

In Greek cosmology, Phobos was one of the two orbiting "steeds", seen pulling the chariot of Mars across the cosmos. In Greek, "phobos" meant fear, or a phobia to flee to some safe place from the approaching Ares, and its bolts of cosmic lightning. (Many hid in caves, cellars or fox holes).

Phobos orbits Mars in only 7 hours, 39 minutes. The spin rate of Mars is 24 hours, 37 minutes. Therefore, to a robot on Mars, Phobos would rise on the western horizon and set in the east. This strange condition occurs nowhere else in the Solar System. Further comment on this condition is reserved for chapter 12.

However, if Mars has always orbited 35,000,000 miles or more from the Earth, as gradualists assume, then why did the Greeks claim to see this tiny satellite? And its partner, littler Deimos?

Evidence # 8 - The Pitted Surface Of Tiny Phobos

The surface of tiny Phobos was photographed by cameras on Mariner 4, 6 and 7. It was in the years 1964 and 1969. Those photos showed Phobos has a series of long, linear grooves, one about 1500 feet wide. Seemingly, another fragment going the same direction skidded across its surface, scuffing and scraping it all the way.

More significantly, the surface of tiny Phobos contains literally hundreds of tiny pitlets and craterlets. The pitlets have been photographed down to a yard or two in diameter. The largest pitlet, Stickney, is 6+ miles in diameter on such a tiny fragment. When that collision occurred, it nearly broke Phobos in half.

Why does Phobos have any pitlets at all? And why so many, and at such a density? The only logical reason is that Mars once had a dense, narrow ring system, a cloud of debris circling its equatorial bulge zone - an extension of its equatorial plane.

The pitlets and craterlets on the surface of Phobos are about as good evidence of a former ring system. Having such evidence is as much as a planetary catastrophists could hope for, save for pitlets on Deimos also.

Judging by the density of craterlets on tiny Phobos, the former ring system must have been a well-populated cloud, rather thick, composed of tens of thousands - no - hundreds of thousands of tiny bits of rocky debris. Figures 10 and 11 portray the pitted and poxed surfaces of both Phobos and Deimos. Figure 12 illustrates the four sweep sites in the catastrophic orbit of Mars.

Figure 11 also illustrates a round-like, semi-smooth surface of Phobos, broken only by Stickney, the largest crater, and by the pitlets and craterlets. Its smooth surface suggests that probably it was formed from material that was from the hot mantle deep under the crust of Astra. Were it were more jagged, it would suggest a crustal fragment.

Figure 10 & 11 - Pitlets on the Surface of Deimos and Phobos

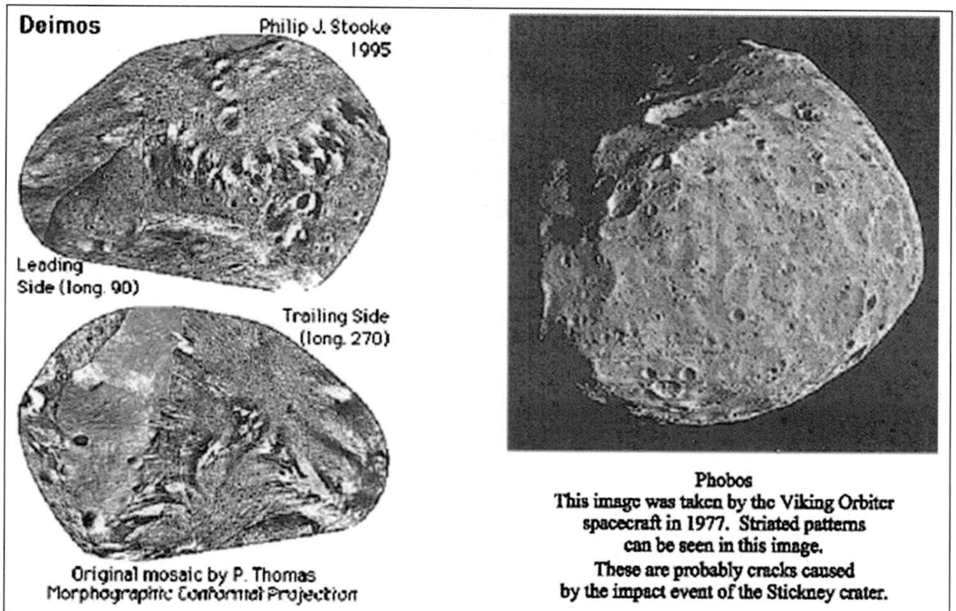

Phobos
This image was taken by the Viking Orbiter
spacecraft in 1977. Striated patterns
can be seen in this image.
These are probably cracks caused
by the impact event of the Stickney crater.

Pitlets on the Surfaces of Deimos and Phobos

Figure 19

Figure 12 - The Catastrophic Third Orbit of Mars and the Four Sweep Zones

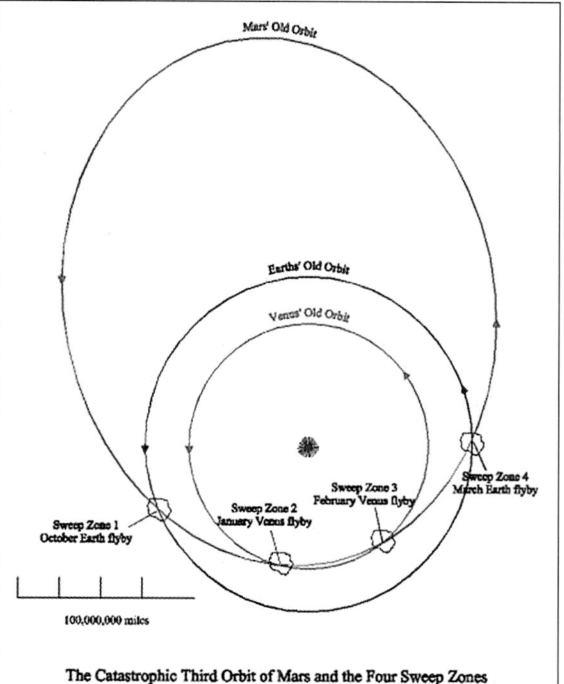

The Catastrophic Third Orbit of Mars and the Four Sweep Zones

That evident location of Phobos, from deep within Astra, may be the precise reason why Phobos is so close in, and never escaped from Mars. Perhaps the energy Phobos suddenly gained in one direction from the explosion canceled the energy of Astra's orbit in another direction, neutralizing Astra's forward motion.

If so, Phobos was left without enough energy to escape Mars, and without enough energy to begin a virgin orbit around the Sun. Understanding this will be helpful in evaluating Evidence # 12 later in this chapter.

Evidence # 9 - The Orbit Of Even Tinier Deimos

Like Phobos, Deimos is also irregularly shaped. It measures 6.5 miles x 7.5 x 10. Deimos also is a rather rounded fragment, as if it too came from the hot, central region of Astra. This also is to be noted when Evidence # 12 is discussed.

Like Phobos, Deimos also is pitted and poxed with hundreds of tiny craterlets, down to a yard or two in diameter. Thus, Deimos and Phobos have both experienced going through a cloud of ring system debris, and have done so repeatedly. This is the second, example of a poxed asteroid from the former ring system of Mars, but not the last.

The orbit of Deimos is 14,500 miles from the center of Mars, and is 12,400 miles from its severely-scarred surface. Its orbit is less than seven radii from the center of Mars. Orbiting Mars, Deimos has a period of 30 hours, 29 minutes and some 45 seconds. The fact that both are poxed indicates that the cloud of debris, once circling Mars, was substantial in numbers of debris.

What is amazing is that were gradualism true, and had Mars has always been 30,000,000 miles or more distant from the Earth, how could the Greeks possibly have seen little Deimos? A body averaging 8 miles in diameter, essentially Deimos is dark in color, with a low reflectance.

For difficulty in seeing, consider Saturn. Its diameter is 71,500 miles. Its average distance is 880,000,000 miles. Its sunlight is one ninetieth as bright as is sunlight on the Earth. Saturn's diameter is one part in 1,230 of its distance. The diameters of Deimos average 8 miles. Our opinion is that Deimos could be seen when it was twice as far away as the Moon. This means it was visible for 20 to 30 hours during Mars flybys. Phobos was visible for even longer.

Moreover, why did the Greeks name it for a word that means another aspect of great fear, or dread? In addition, why did the Greeks correctly describe it as circling Mars, like a wheel of a celestial chariot? Gradualists usually clear their throat, and shift the subject of discussion away from questions like this. But why not face issues like this openly. What does playing ostrich solve? Nothing. Once again, how could the Greeks have named

Deimos and Phobos if they couldn't see them?

Deimos orbits Mars today for two reasons. One is because, as a virgin fragment, its orbit velocity was apparently canceled by the vector of the explosion, in the opposite direction. The second reason is that it was not quite swept away from Mars by the Earth or Venus during the ancient flybys.

Evidence # 10 - Pitlets On The Surface Of Deimos

Figure 10 also portrays the pitlets on tiny Deimos. This means the pitlets on Phobos were not a fluke. This pair of twin conditions can be explained by both repeatedly passing through a cloud of debris such as a ring system. No other logical explanation comes to mind.

While the orbit of Deimos is not as close to the Martian Roche Limit as is that of Phobos, still it is close, within 15,000 miles. Its pitlets are the tenth category of evidence for the fragmentation of Astra on the Mars Roche Limit. Its pitlets are an additional bit of evidence for the Catastrophic Third Orbit of Mars, an orbit of .56 or so in eccentricity, which orbited out to the region of Astra, yet also orbited inward, inside the orbits of Venus and the Earth.

Like Phobos, Deimos also was heavily battered and bombarded with tiny debris. Its numerous pitlets have been photographed by the Mariner missions. Pitlets on Deimos were photographed down to a yard in diameter; the count of pitlets on tiny Deimos also runs into the hundreds.

> Deimos differs from Phobos in certain important respects. There are no craters larger than 3 km [1.9 miles] in diameter. The two most prominent craters have been named Swift and Voltaire, but all in all, craters are much more subdued on Deimos than on Phobos. Most are flat floored, with distinct breaks of the walls. It is interesting that neither satellite shows any trace of craters with central peaks or terraced walls. [n1]

The pitlets and craterlets on Deimos indicate that it, too, was banged up by debris, and just as savagely as was Phobos. Why is Deimos 9,000 miles farther out than Phobos? Perhaps little Deimos was almost swept out during one of the Earth or Venus flybys. "Almost swept out" implies that other ring system debris was swept out.

Magicians can make handkerchiefs and rabbits vanish at will. Gradualists affirm that asteroid belts, planets replete with spin rates and satellites all could appear anywhere, "if given enough time." After all, given enough time, can't anything happen? What has 20th century cosmology been so far, if it hasn't been one big magic show?

What, then, caused the ring system of Mars to vanish? Mars-Earth skirmishes. Mars-Venus skirmishes. The Barringer Crater, a recent crater in Northern Arizona, is the result of a chunk of debris hitting the Earth. This fresh crater is 570 feet deep, and 4,150 feet in diameter and is located 40 miles east of Flagstaff. It happened within the collective memory of the Hopi tribe of Arizona.

The Mars-Earth Wars And The Mars-Venus Wars

The answer to the question about the swept out ring system of Mars is illustrated by Figure 12. Figure 12 portrays the Third Orbit of Mars. This catastrophic orbit seemingly was both pre-Astra and post-Astra. This model presumes that Jupiter, in 1:6 orbital resonance with Mars, corrected any modest shifts in the orbit of Mars caused by the fragments of Astra. The Greeks presumed that father Zeus, Jupiter, was the choreographer of the cosmos.

The foundation for the Catastrophic Third Orbit of Mars, is laid in a statistical format in Chapters 9 and 10. Clues to its veracity occur in Chapters 11 and 12. The key topic in those chapters is, "How did Mars get from its catastrophic to its radically different, much more circular modern orbit?" One important issue is about shifts in energy involving Mars, Venus and the Earth. A second, equally important issue is about shifts in angular momentum simultaneously.

One concern is about how many Mars-Earth Wars occurred, and another is how many Mars-Venus Wars also occurred. How close did Mars come to Venus, and to the Earth. These issues will be addressed also. (One thing is addressed at a time.)

For the present discussion, the model holds that there were over 150 Mars-Earth skirmishes. They ranged from as distant as 75,000-mile flybys to one, the closest, as near as 15,000 miles, measured planet center to center.

There is no way to lay a foundation for the count of Mars-Venus Wars. Probably, they were as numerous as were the Mars-Earth Wars. The scarred surface of Venus attests to repeated Mars-Venus skirmishes, with some of them having been all out battles.

Evidence # 11 - The Pitlets On Gaspra

Gaspra is a small inner asteroid, like Adonis, Amor, Apollo, Eros, Geographos, Hermes and Icarus. Gaspra has an orbit that approaches the Earth's orbit at the 182nd longitude. This is the Earth's position of March 23. The obvious call is that Gaspra was a Mars asteroid that was swept out of the Martian ring system during one of the March Passovers, ancient March Mars flybys.

Gaspra, like Deimos and Phobos, is heavily pitted with craterlets. Its numerous pitlets substantiate an experience much like Deimos and Phobos, and for almost as long. Figure 13 illustrates the pitlets poxing the asteroid Gaspra; compare it with the pitlets in Figures 10 and 11.

PREDICTION VI. It is predicted that when the surfaces of other inner asteroids are photographed, they also will be found with poxed surfaces. The subject includes such inner asteroids as Icarus, Amor, Apollo, Chiron, Moore, Eros, Adonis, Geographos and Hermes.

Figure 13 - The Pitlets of Gaspra

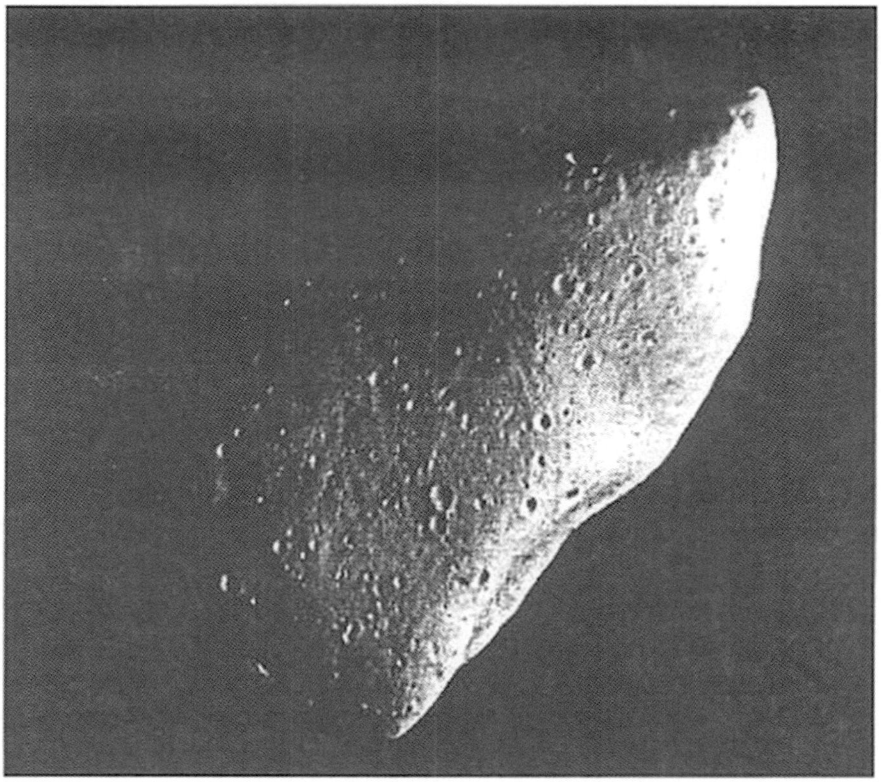

Gaspra is an irregular body with dimensions about 19 X 12 X 11 km. It is similar in size to Phobos. The large Concavity on the lower right limb is about 6 km across, the prominent crater on the terminator about 1.5 km. A striking feature of Gaspra's surface is the abundance of small craters. More than 600 craters, 100-500 meters in diameter are visible here. Gaspra's very irregular shape suggests that the asteroid was derived from a larger body by catastrophic collisions.

PREDICTION VII. It is also predicted that upon closer surveying of the region of the orbit of Phobos, there will be found a small collection of orbiting debris. When found, this debris will be remnants of the former ring system that also, like Deimos and Phobos, were not swept out during the planet wars of Mars.

Since no astronomers yet concur that Mars once had a former ring system, they would have no reason to make a minute inspection of the region of Phobos-Deimos in the space surrounding Mars. Indeed an unconfirmed report has come in from Van Flandern that one of the recent Soviet missions to Mars indeed encountered a thin ring system near the Roche Limit of Mars. This unconfirmed report was received after our prediction was made, which was shared with astronomer Van Flandern.

In the 1920's, a trans-Neptunian planet was suspected, due to minute, unexplained perturbations Neptune experienced. A ten-year search program was instituted. In due time, after comparisons of tens of thousands of photographic plates, Tombaugh discovered Pluto in 1930.

Another search program needs to be instituted in the late 1990's to discover and assess the remains of the ancient ring system of Mars. It need not be a ten-year search program; it could be done with the next Mars mission. This model predicts not only that the remnants of that former ring system will be found. It also implies that the tour directors for Mars missions should be careful; encountering such debris in a careless entry path could destroy their multi-billion dollar project.

Astronomers so far have trouble understanding that the genesis of the asteroids is a fragmentation on the Roche Limit of Mars. It is an obvious premise, and a simple one, with, so far, eleven categories of evidence. When they become suspicious, a search program will be instituted. It will be successful. But ring system debris will be small, and being small, it likely will not be found by accident. This model addresses the origin, the travails and the travels of the fragments of Astra. The pitlets on the surfaces of Deimos, Phobos and Gaspra are part of that travail.

Evidence # 12 - Traditionalists And Their Lack Of Satisfactory Theory On Asteroid Genesis

In a mystery, sometimes there are ten possible answers, or suspects. If nine of them can be eliminated, the tenth explanation, however unlikely at the first glance, becomes the probable culprit after full analysis. So said the master sleuth, Sherlock Holmes.

THE VAN FLANDERN APPROACH TO ASTEROID GENESIS. Tom Van Flandern is a well-educated, well-connected, contemporary astronomer. One of his jobs includes reviewing manuscripts in astronomy for publications. He is courageous, honest, poised, pleasant and sincere. He classifies himself not as a gradualist, but rather as an "asteroid catastrophist." Astronomers have been frustrated now for a century in endeavoring to explain the genesis of the asteroids. That their attempts have been unsatisfactory is acknowledged by Van Flandern's recent attempt for a new explanation. His is an attempt with a new pair of wrinkles.

In the past, the explanation usually has been by "slow accretion." The time frame has been 4.6 billion years, +/-. a few hundred million. The source of the asteroid material has been a hiccup or two from the Sun, the cause of the hiccups being unknown. Such has been the accepted menu for 150 years; this menu lacks only two things, evidence and logic.

First, asteroids in space tend to disperse. They do not tend to condense or "accrete." Miscellaneous dust or debris does the same in all parts of the Solar System. To date, perhaps in just 10,000 years, the asteroid perihelion's have dispersed, and spread out to a belt 40 million miles deep and 80,000,000 miles wide.

After discussing the four inner planets, Mars being last, Van Flandern follows the Canadian astronomer Michael Ovenden's novel idea with considerable enthusiasm. Ovenden claims there used to been a huge, gaseous planet this side of Jupiter. It was 90% of the mass of Saturn, Saturn being 95 Earth masses. Thus it was 80 or 85 Earth masses.

> If so, its former parent planet would have been the next plant out from the Sun. This would seem to be the location of the "missing planet" where the asteroid belt is now. In the chapter "Do Planets Explode?" we discuss evidence that a large planet, perhaps Saturn-sized, orbited there until recently in the solar system. Let us refer to this missing planet as "Planet K," following the lead of Canadian astronomer Michael Ovenden, who argued that the appropriate name for this hypothetical exploded body, taken from modern mythology, would be "Krypton." ... Superman's home planet, of course. [n2]

In an astronomical convention in Europe, Van Flandern advocated Ovenden's idea, and was laughed at and humiliated. He was even denied the opportunity to respond to his critics. Subsequently Van Flandern has changed his mind on some aspects of Ovenden's idea, but persists with others. This is an indication of his grit and courage, of his commitment and of his general dissatisfaction with the "traditional menu."

Van Flandern now supposes that this region was inhabited by a planet much smaller than Saturn, but like Ovenden, he feels that missing planet "novaed" - it burst into a massively bright and temporary star. He readjusted his view that the former planet, Krypton, may have been only 4,000 miles in diameter - Mars sized and not Saturn-sized.

One of the suspect ideas of the Ovenden and Van Flandern approach is the "nova" part. Some novas are periodic, recurring periodically, perhaps with periods of 40 or 60 years.

Most astronomers, viewing novas elsewhere in the galaxy, NOT SEEING THE CLOSE DETAILS, profess to not know why novas occur.

> NOVA. ... Causes. The causes of these outbursts are still unknown. In the case of ordinary novas, the star for some reason blows off its outer layers, but the amounts of energy and matter involved are not sufficient to produce any far reaching change in the star. There is some evidence suggesting that ordinary novas are members of binary star systems. [n3]

A case was made in "The Recent Organization of the Solar System," our Volume I, as to what causes novas. Evidence on four or five levels was presented indicating that the Sun had novaed, and recently as astronomers measure time. It is still in the process of cooling and shrinking, at a rate of 120 feet per day. Your authors claim to understand what caused the Sun to have a nova.

If our evidence and analysis is sound, it can be said that neither Ovenden nor Van Flandern understand the genesis of novas. They alleged that the nova exploded; novas do not explode. They merely rise rapidly to a maximum brilliancy in 40 or 50 hours. They suspect the debris of the nova is the genesis of the asteroids. They equate a nova with an explosion, which a nova is not.

Therefore, their assumptions are flawed, and their ideas are suspect as explanations for both the genesis of the asteroids and for the presence of little Deimos and Phobos, orbiting Mars.

For example, they interchange the ideas of "an explosion" and a nova. Novas typically flare from invisibility to maximum luminosity in 40 to 50 hours. Then they gradually diminish in luminosity for six months, perhaps up to 24 months, when they become invisible.

The evidence assessed in Volume I indicates novas are generated by sudden, immense tides in stars, by close flybys; they have nothing to do with explosions or fragmentation's. Fragmentation's are produced by penetrations of Roche Limits. To reiterate, a suddenly expanding nova, which occurs in 34 or 40 hours, is not an explosion, which occurs in seconds.

Another aspect of the Ovenden-Van Flandern approach is bothersome. It is what caused the fragmentation of the asteroids. In the case of this mystery, the physical geography of the surface of Mars virtually begs to convict the Martian Roche Limit. The red planet has over 90% of its craters on one side.

Mars today is a neighbor to the asteroid belt. And if our model is correct, it formerly orbited in the asteroid belt - before there was an asteroid belt. Ovenden and Van Flandern have missed the obvious. They both are good astronomers, but cannot be complimented as cosmologists. To repeat, cosmology is a study of the history of the Solar System, astronomy is a study of its present arrangement.

Evidence # 13 - The Genesis of The Orbits of Deimos and Phobos

From here, Van Flandern's saga becomes even more improbable. The subject is the orbits of the two little satellites of Mars, Deimos and Phobos. What traditional gradualist astronomy badly needs is a good explanation of how Mars captured them.

The traditional menu for the last 100 years has been that somehow, little Mars caused an asteroid or two to slow down on the fly from velocities close to 50,000 mph down to 3,000 or 4,000 mph. Simultaneously, Mars redirected it or them into a steep turn to circle itself. Such a capture on the fly is beyond unlikely. The physics of such a slowdown and sudden turn are impossible, as astronomers readily acknowledge. Only magicians can accomplish such things, but that is on stage and with mirrors and trap doors. Yet astronomers have ignored the veto by physics, and have affirmed that such a capture on the fly must have happened.

Their reason is that little Deimos and Phobos are there, so it must have happened that way, however unlikely. Circular reasoning. Moreover, such an unlikely scenario happened not once, but twice, once for each satellite captured on the fly. This is impossibility squared; it is unworthy of serious discussion.

Privately, astronomers, when discussing the capture of Deimos and Phobos on the fly, smile a little nervously and quickly change the subject. Mathematicians are aghast. Physicists weep. Engineers laugh. None take that explanation very seriously because all realize a significant slowing down of an asteroid's velocity by little Mars is not possible even once. In addition, a redirection into a tight capture orbit is equally illogical ... impossible.

Van Flandern understands correctly that there is a need for better theory for Mars to capture Deimos and Phobos on the fly. Bravely, and creatively, he addresses a capture by Mars of Deimos and Phobos in another way.

> The Martian moons Phobos and Deimos appear to be captured asteroids. [n4]
>
> Among the many such collisions likely to occur, some may be suitable to leave a few of the larger masses gravitationally bound to Mars in a permanent way. Moreover, continued collision interaction with the temporary bodies in orbit will tend to drive the larger captured satellites into somewhat more circular and somewhat more equatorial orbits, ... [n5]

Van Flandern proposes there was a cloud of asteroids that approached Mars. He thinks two or more asteroids collided in space near Mars, decelerating. Twice it happened. But once again, there are troubling problems in his details.

1. DIRECTIONS OF MOTION. In order to collide in this prescribed manner, one asteroid would better revolve in the clockwise motion around the Sun and the other in the counter-clockwise direction. However all asteroids revolve in the counter-clockwise manner.

2. ONE DIRECTION OF REVOLVING. If asteroids were to collide in space at approximately 50,000 mph, all with prograde motion, they would ricochet and bounce away from each other but would not decelerate appreciably.

3. MARS NOW IS IN AN ASTEROID-FREE ZONE. The modern orbit of Mars averages 141,600,000 miles from the Sun; it orbits in a zone between 128,400,000 miles and 154,900,000 miles. There are only three asteroids orbiting in this zone to our knowledge, Icarus, Apollo and Amor. All appear to have been sweepings from Mars' once well-populated ring system.

Troubling Issue # 1. Van Flandern proposes that a cloud of asteroids swept in near Mars and from that cloud, two or more collided, decelerated and were captured. But there is no evidence any cloud of asteroids ever being near the modern orbit of Mars. His is a postulation without any proper foundation of evidence.

Troubling Issue # 2. In the Van Flandern approach, the orbits of two captive asteroids should have "apogees" at much greater distances than they do. The "apogee" of Phobos is under 6,000 miles from the center of Mars, and that of Deimos is under 16,000 miles, both very close in, too close, too near to Mars. What they are near is the Mars Roche Limit.

Troubling Issue # 3. The surfaces of both Deimos and Phobos have hundreds of craterlets, or pitlets. Craterlets and pitlets require nearby debris and in quantity. Van Flandern's postulation does not address the pitted physical geographies of the surfaces of either Deimos and Phobos.

Troubling Issue # 4. Greek cosmo-mythology includes the story of Astra and her disappearance in the Solar System. True, the tradition must have come from Hittite, Indo-European and Sumerian ancestors. Even so, where is the time and the resources for the pitting of Deimos and Phobos? This model, on the other hand, addresses each of these circumstances with a foundation of both evidence and logic.

The four nearest asteroids to the modern orbit of Mars of which we are aware are Bamberga, at 165,000,000 miles, Iris at 171,000,000, Flora at 173,000,000 miles and Hebe at 179,500,000 miles from the Sun.

The next 5,000 are farther out, and over 75% of them are well beyond 200,000,000 miles. They are 50 million miles distant from the modern orbit of Mars. Van Flandern overlooks both (a) the paucity of asteroids in the region where Mars now roams and (b) the abundance of asteroids in the region where Mars once roamed.

Van Flandern, an editor of articles in astronomy, is well positioned as an astronomer, but not as a cosmologist. He pays no attention to Greek cosmology, to ancient Roman literature, to ancient Hebrew literature, to Vedic traditions and literature, to ancient Chinese cosmology, or ancient American Indian lore involving cosmic traditions.

The simple and logical answer is that Mars captured a ring of asteroids when Astra fragmented. They were the less energetic fragments in the aftermath of the explosion. Deimos and Phobos somehow survived the Mars-Earth Wars and the Mars-Venus Wars. If these two could survive, other tiny debris probably did also, and if so, the debris is small or it would already have been seen, but it can be found if and when a search program is instituted.

The truth is that traditionalists in astronomy have had more than a century to come up with a logical explanation with evidence for the acquisition of, and the retention of two tiny moons by Mars, Deimos and Phobos. Van Flandern's recent and courageous endeavor (1993) is best viewed as an attempt to rectify that condition. But the overview is that his endeavor still falls far short. His attempt is best viewed as merely the latest salute to 100 years of failure in gradualist cosmology.

As is mentioned above, the explanation for Deimos and Phobos orbiting Mars is simple: they are fragments of Astra that became Mars asteroids. And they remained Mars asteroids throughout the Mars-Earth Wars era. Hence the ancient Greeks have accounts of them, the "steeds of Ares," accompanying Ares, the "bane of mortals." The blind bard Homer some 2860 years ago was both profound and precisely correct.

Very appropriately, Van Flandern registers dissatisfaction with tradition explanations. His extensive research on the subject of Mars and its two moons, is encouraging. His willingness to entertain non-traditional ideas (up to a point) also is encouraging. Perhaps, without realizing it, Van Flandern is already 25% of the way to becoming a planetary catastrophist - a Mars planetary catastrophist.

> *Cease, my son, to hear the instruction that causeth to err from the words of knowledge.* - Solomon (Proverbs 26:12.)

Conclusion

There is a logic in the belief that once Astra fragmented, and a small percent, possibly as little as 1% or 2% of its debris orbited Mars rather than the Sun.

There is ample evidence that Mars once had a thin ring system - all rings are thin - and it was well-populated. Cratered Deimos and poxed Phobos are vestiges of that former ring system as is poxed Gaspra.

Story 12 is a logical explanation for THE POXING OF DEIMOS AND PHOBOS.

Story 13 is the explanation for THE POXING OF GASPRA, AN INNER ASTEROID, once in the ring system of Mars.. Other suspects for being poxed like Deimos, Phobos and Gaspra are the inner asteroids Adonis, Amor, Apollo, Eros, Geographos, Hermes and Icarus.

Story 14 is the explanation for HOW MARS ACQUIRED ITS TWO TINY SATELLITES, DEIMOS AND PHOBOS. They were not captured on the fly, at some 50,000 mph. They were captured as low energy fragments from the explosion of Astra, at which time they began to orbit Mars. This may have been less than 12,000 years ago.

A search program for remaining ring system debris should be instituted. A count program for the expected density of craterlets on the other inner asteroids (Gaspra included) should be planned.

Some inner asteroids were swept out by Venus. Some were swept out by the Earth during March Passover flybys. Some were swept out by the Earth during October flybys. The Barringer Crater in Arizona is one of several examples that some substantial debris from the ring system of Mars hit the Earth's terrestrial surface.

Perhaps an analysis of the orbits of the eight or ten inner asteroids can determine which kind of a flyby it was when each was swept out.

It may be alleged by critics and skeptics that in this model of planetary catastrophism, planets were flying about all around the Solar System. NOT SO. NO CHANGES occurred to the orbits of Mercury, Saturn, Uranus, Neptune or Pluto. Jupiter's orbit was only tickled and it captured a mere handful of asteroids. (See Table XIII). Some minor changes occurred to the orbits of the Moon and Venus. Major changes occurred only to the orbits of Astra, the Earth and Mars.

During the closest flyby of all, Mars almost became fragments (like Astra) as it approached as close as 15,000 miles to the Earth during Noah's Flood. This was only 4,000 miles from the Earth's Roche Limit. This flyby, in 2484 B.C.E., was the closest of all. It is to be described and analyzed in detail in Volume III, entitled The Flood of Noah. Mars survived that occasion, but just barely; according to a baker's dozen of ancient reports, Mr. and Mrs. Noah also survived, also just barely.

With story 14, the reader now is 44% of the way to the penthouse; the scenic view keeps expanding.

CHAPTER 7

The Flood of Mars - Followed By Its Ice Age

> Arrhenius remarked concerning those who developed the law of the indestructibility of energy: "It is very significant that not one of these men was a scientist by profession." A draper first saw bacteria under a microscope. A Unitarian minister first isolated oxygen. Mendel, a monastic school teacher, discovered the laws of heredity. In mathematics, "analytical geometry was the independent invention to two men - Fermi, a lawyer; and Descartes, a philosopher - neither of whom was a professional mathematician.

Introduction

According to the Earth's stratigraphic record, of both the Eastern and the Western Hemispheres, it has experienced numerous and repeated floods. The flood in the time of Noah was the most remarkable event to the ancients of the Eastern Hemisphere, but it was by no means the only celestial event.

The Earth also has experienced at least one ice age - possibly two or, less likely, three according to planetary catastrophism, but five or six according to gradualists. But what about the surface of our neighbor, Mars?

The atmosphere of Mars is unlike the Earth's in all ways. Mars has very little atmosphere. At its surface, atmospheric pressure is less than 1% of the Earth's pressure. Further, of what thin atmosphere Mars has, it contains only tiny percentages of water vapor (0.03%) or oxygen (0.13%). There are no oceans, lakes or seas on Mars. Therefore, how likely is it that the surface of Mars should show evidence of water activity? Of rampaging water action?

Perhaps the most controversial of all Martian surface features are the channels. Did they form as a result of the action of running water or not? ...

It is possible to recognize three kinds of channel; (sic) runoff channels, outflow channels and fretted channels. [n1]

In viewing the early photos of the surface of Mars from Mariner 4, 6 and 7 in the 1960's, astronomers and geologists were astounded at the seemingly impossible. Numerous dry river beds were found, some once containing water flowing at speeds at velocities estimated from 25 mph to 40 mph.

This is on a planet whose normal surface night time temperatures approach -200° F. The longest of the Martian dry river beds is 400 miles, which is long enough for 16 hours of rampaging waters, flash flooding at 25 mph. It would seem that Mars had rampaging rivers, but those rivers did not rampage for as long as two 24-hour days. The waters soon froze in the coldness of their first Martian night.

All of its dry river beds but one are in what has been designated as the Eastern Hemisphere of Mars, 0° to longitude 180 W. The one dry river bed which is in the Western Hemisphere is at its extreme eastern edge.

Thus, like its asteroid craters, its dry river beds have a one-hemisphere preference, in this case its Eastern Hemisphere. It was found that the craters of Mars have a one-hemisphere preference; now its dry river beds do also. However, the two hemispheres do not share the same centers either in latitude or longitude.

It was found that the asteroid fragments hitting Mars arrived suddenly. Since the rivers have a one-hemisphere preference, and since they rampaged, but only for a dozen or so hours, it is concluded that the waters of Mars, like the fragments of Astra, arrived suddenly, from space.

Water, in immense volumes (like a flash flood) once flowed on the surface of a planet with no water clouds, no oceans, no rain, no rivers, no canals and no snow falls - at least from water-based snow. (Mars does have some carbon dioxide snow and concentrated carbon dioxide ice in its polar regions.)

The Rarefied Atmosphere Of Mars

There is ample evidence on Mars of the work of a fluid agent. There are numerous examples of what appear to be ancient stream beds and other water-sculptured terrain. [n2]

The atmospheric pressure on Mars is less than one-hundredth of what it is on Earth. In effect, this means the air is as rarefied on the Martian surface as it is at the height of 30,000 m on Earth. At pressures such as this and at Martian temperatures, liquid water becomes unstable and freezes on the surface.

Another effect of such a rarefied atmosphere is that the transportation of material along the surface (saltation) and the raising of fine material to form dust clouds are a conspicuous characteristic of the Martian landscape. [n3]

Water, like dry ice, requires pressure to keep from effervescing. In addition to being a rarefied atmosphere on Mars, that thin atmosphere today contains virtually no water vapor - or oxygen - or nitrogen. By percentages, its rarefied atmosphere has the following composition. In an already rarefied atmosphere, .03% is three parts per 10,000.

Table VIII - Makeup of Mars' Atmosphere

Carbon dioxide	95.3%		Carbon monoxide	00.07%
Nitrogen	02.7		WATER VAPOR	00.03
Argon	01.6		Neon	00.00025
Oxygen	00.1		Krypton	00.00003

Its surface temperatures almost always are subzero, often far subzero. Occasionally, in equatorial zones, in its summertime, in the middle of its 24-hour day, surface temperatures may rise to 50° F., some 20° above freezing, for a few hours. But when the middle of the night comes, temperatures again plunge to -150(or -180° F., even in summer time.

On the Earth at night time, due to radiation, with no wind, temperatures may drop as much as 3° per hour. Temperatures during a Martian night can drop from +50° F. to -190° F. in twelve hours. This 240° in twelve hours is an average drop of 20° F. per hour during the Martian night. What an unlikely place is the surface of Mars to expect evidence of onrushing rivers, sudden lakes or as some have said, water-carrying canals, involving life and civilization on Mars?

Among planets and their atmospheres, there is a phenomenon known as "escape velocity." It relates to the energy of motion of various molecules. Different gases have different escape velocities, with lighter gases having lower escape velocities.

The more massive a planet is, the more easily its gravity retains all molecules that can comprise an atmosphere. Conversely, the smaller a planet is, the harder it is for it to retain the lighter gases that could compose an atmosphere.

Table IX - The Escape Velocities For The Nearby Planets

Jupiter	59.6		Mars	5.02
Earth	11.18		Mercury	4.25
Venus	10.35		Moon	2.37

On small planets, due to molecular energy of the lighter gases, they tend to just float off into space. On larger planets, all gaseous molecules are retained. The Moon is too small in mass to retain an atmosphere; so is Mercury. Venus and the Earth are massive enough, and readily retain atmospheres. Mars, one-ninth of the Earth's mass, and nine times the mass of the Moon, is below middling in mass in terms of ability to retain an atmosphere..

The mass of Mars is effective in retaining certain heavier gases such as carbon dioxide, argon, neon and krypton, but not oxygen, water vapor or nitrogen. Hence, as cited, Mars has a rarefied atmosphere, one comprised 97% of the heavier gases - 95.3% carbon dioxide, 1.6% argon.

The small mass of Mars is the reason Mars has so little atmosphere, but there is some. Mars neither presently has nor ever has had much oxygen or water vapor in its atmosphere. In fact by Earth or Venus standards, MARS IS EXTREMELY DRY, AND HAS NEVER HAD MUCH ATMOSPHERE AT ALL. All astronomers and most physicists realize this.

Therefore, when scientists suggest that there "might be" or "might have been" fossils of ancient life on Mars, it is contrary to the logic of escape velocity for O2, N2 and H2O. It is also contrary to the logic in the photographic evidence from the Mariner 6, 7 and 9 missions.

Some scientists who so speculate are merely posturing, needing some basis for favorable project grants from legislative bodies that are unaware of these troubling realities. Yet, such allegations do help to convince Congress to appropriate billions of dollars for Mars projects. Such allegations, if in concert with a supportive press, does help to drum up support.

However, there are valid reasons for funding missions to Mars. One reason is to search the debris of, reportedly, a former ring system. A second reason is to search its lava outflows for evidence of paleomagnetic polarity reversals.

A third reason is to determine the isotopic nature of any former ice that is locked in subsurface soils. Does Mars ice lack the oxygen 18 isotope and the deuterium mixture that characterizes Antarctic ice, 4,000 feet below sea level, resting on bed rock, today? Or is Martian ice similar in character similar to water and ice on our planet?

Gases in rarefied atmospheres such as that of Mars escape at different rates. Free hydrogen (H2) and helium escape first. Water vapor escapes second, atomic weight 18, H2O. Third fastest is nitrogen, atomic weight 28. Fourth fastest is oxygen, atomic weight 32. These gases cannot be retained by Mars at either its present or its ancient mass.

Argon, atomic weight 40, can be retained by Mars. Carbon dioxide, CO2, atomic weight 44 can be retained also, as can some other exotic gases like neon and krypton. This being said, the surface of nearby Mars is indeed a most unlikely place to find much in the way of usable water, with its deep subzero temperatures at night.

Thus, on Mars there is no growing season, no running water, no irrigation districts, no water-filled canals, no hydraulic engineers, no siphons, no flumes, and no farmers irrigating their fields on the red planet. But what has the literature of astronomy had to say on this before the Mariner 6, 7 and 9 missions, 1969-1971? The fact is that idea of life on Mars has fascinated scientists and non-scientists alike for the last two centuries.

Wishful Thinking By Astronomers For Life On Mars

William Herschel (1738 - 1822)

William Herschel was an amateur astronomer who built the finest reflecting telescope of his time, in his back yard. Using it, he became a professional astronomer. With his telescope, he discovered the planet Uranus, two satellites of Saturn and the rate of rotation of Mars. Herschel hoped that what he was seeing on the surface of Mars was a primordial ocean. He became the first president of the Royal Astronomical Society in London.

By Herschel's time, astronomers, looking at poorly defined images of Mars, were speculating that Mars was a planet that was much like the Earth, perhaps brimming with life. Herschel supported such speculations, adding that on the surface of Mars, dark spots were Martian oceans. Escape velocities for gases at that time were not understood.

But to Herschel, this meant water "must be" abundant on the surface of Mars. He also considered that the polar ice caps that he saw were evidence of water snow and ice; white carbon dioxide ice didn't occur to him.

Herschel, like Swedenborg, was one of the earliest examples of astronomers wanting to believe there is or was life on other planets. The wish for life to be found elsewhere in the Solar System has not vanished with adverse evidence. It continues but is located elsewhere; merely the distances from the Earth have been extended from 150 million miles to other galaxies thousands of light years distant.

In the first half of our century, the popular wish or yearning for this legend supported Buck Rogers and Flash Gordon comic strips and sold countless tabloid newspapers. Planet wars was the theme in a few movies of the 1930's, usually featuring Mars inhabited by men made of clay.

More recently, this yearning has supported some popular movies and television series. Necessarily now, the region for life on another planet is far away in the Milky Way or in another galaxy. No longer is it the nearby planets, the evidence now being so adverse. As has often been the case, wishful thinking runs ahead of evidence.

Giovanni Schiaparelli (1835 - 1910)

Giovanni Schiaparelli was an excellent astronomer, and he rose to be the director of the Brera Observatory in Milan. He made some important discoveries relating to asteroids, comets and meteor streams. He also determined the axis of rotation on Mars, and devoted himself to writing a series of essays on the surface of Mars.

Occasions occur once every 15 years when, in their orbits, Mars and the Earth are particularly close - at distances of 40,000,000 to 45,000,000 miles. The summer of 1877 was such an occasion. The weather was favorable. Schiaparelli had a new, fine 8-inch refractor telescope. He focused it on the surface of Mars. He saw numerous lines, and reported them, in Italian naturally, as "canali."

Schiaparelli's report of "canali," or lines on Mars by telegraph was transmitted to New York. The Italian word "canali" could have been translated equally well as "canals," "channels" or "lines." Schiaparelli meant "lines."

But in boisterous New York City, this most exuberant of cities in l877, as either chance or "excessive optimism" would have it, "canali" was translated as "canals" - with all that canals imply in the English language. Canals imply metering equipment, siphons, ditches, surveys, hydraulic engineers, artificial reservoirs, irrigation, flow gates, flumes, farmers, agriculture, cities, etc.

Schiaparelli's report, as translated in New York, became sensational. In cities up and down the Eastern Seaboard, newspapers containing mistranslations of Schiaparelli's account sold in record numbers. Canals meant life was indeed on another planet, and a neighboring one at that. "Life" included irrigation works, oases, engineers, farmers, perhaps even cities, universities and other astronomers!

Canals also imply intelligent engineering, surveys, major construction projects and an intensive agriculture. Probably Mars had intelligent engineers, brawny farmers, fragrant fields, overflowing oases, and perhaps even avid astronomers with their telescopes, watching our planet.

The theories of Charles Darwin, seemingly, had been right; moreover, confirmation of Darwin's theory now had surfaced in record time, less that thirty years. The speculation developed in a carnival atmosphere.

Percival Lowell (1855 - 1916)

Lowell, a member of a brilliant, New England family, was educated at Harvard but not in science. His studies were in business, literature and the arts. He spent several years in the Far East, including Korea and Japan. There, he had written "Choson" (Korea) (1885), "The Soul of the Far East" (1888) and "Occult Japan" (1895).

In the mid 1890's, inspired by the translation of Schiaparelli's "canals" on Mars, Lowell determined to devote his energy and his inheritance to this popular matter. He took up the pen and went on a well paid lecture circuit. He wrote "Mars and its Canals" (1906), "Mars as the Abode of Life" (1908), "The Evolution of the Worlds" (1910) and in the year he died, "The Genesis of the Planets" (1916).

Lowell became a popular public lecturer and received numerous scientific honors. He associated with astronomers. He was instrumental in the choice of a site for a new astronomical observatory, near Flagstaff, Arizona. It was over 7,000 feet above sea level. There, skies were usually clear at night, and there were no glaring city lights. The atmosphere was thinner, and there was no smog. Now famous in science, and a leading proponent for this new observatory, he was honored by having this new observatory named in his honor, the Lowell Observatory.

Lowell, among others, came to suspect the existence of at least one Trans-Neptunian planet. Tiny perturbations in the position of Neptune pointed in that direction. At this observatory, a fifteen-year search project was instituted to find one. Fittingly, it was at this observatory where the planet Pluto was discovered in 1930 by Clyde Tombaugh.

Tombaugh was accorded the privilege of naming the new planet; he chose the Greek name "Pluto." Pluto was the Greek deity of the underworld. And not coincidentally, its first two letters, "P" and "L", were the initials of his mentor, and his observatory's founder, Percival Lowell. It was all quite fitting.

Lowell inspired much new and burgeoning interest in astronomy. His books inspired comic strips of the 1920's and B-rated cinemas of men on other planets in the 1930's and 1940's. In his wake, Flash Gordon and Buck Rogers became household words. More recently, this widespread yearning, partly from his heritage, has inspired Star Trek and other television programs. But they are about intergalactic travel, not Martian adventures.

In the years following Lowell's death (1916), every time there was an "opposition" (near approach) of Mars, once every 15 years, telescopes would be turned toward Mars, with the big question being asked, "Is there life on Mars?" The telescopes were not good enough to resolve the question. About 50% of the astronomers said they didn't see lines on Mars, and the other half affirmed that lines seemed to be there.

Sketches of those legendary lines were made by astronomers at various locations, independently. Next they were collected and were compared. Collusion between sketchers was out of the question. When some of those sketches more or less matched, they were used by the "Life on Mars" proponents to announce that here was additional proof; indeed, there was life on Mars.

The non-matching and blank sketches were ignored. The astronomers who didn't see lines on Mars, and were more skeptical, also were less vocal, or else they weren't nearly as good copy for the newspapers. And so the issue continued, basically unresolved, after Lowell's death for another 48 years.

The Pickerings

Edward Charles Pickering (1846-1919) was a famous American astronomer of the early 20th century, among other positions, was director of the southern station of Harvard College Observatory, at Arequipa, Peru. His brother, William Henry Pickering (1858-1938) cooperated with him, and discovered another satellite of Saturn and predicted the ninth planet, Pluto.

> He made careful studies of Mars, claiming that the canals crossed the dark areas as well as the other tracts, and that where two canals crossed there was usually a dark, circular path or "oasis." Eventually he proved that the dark areas on Mars were not aqueous. [n4]

James S. Pickering wrote about life on Mars as late as 1959 as follows:

> What is the green on Mars? Probably vegetation. There is a seasonal change in the color of the green regions in general, which follows to some extent the pattern of the seasonal color change in vegetation on the Earth. The green areas on Mars are faint and delicate during the Martian spring; they grow much darker and, in some cases, almost blue-green in the Martian summer; yellow and brown in the fall, and a rather lifeless gray in winter. The green regions are probably not covered with grass or trees, because they do not show the same sort of reflection of light that is produced by a chlorophyll plant. The vegetation, if such it is, is likely to be a sort of lichen which can grow in little or no soil, with a minimum of moisture and can withstand extremely cold temperatures. [n5]

So the issue continued, probable according to many authorities, but unresolved, ... until the 1960's.

Photos Of The Surface Of Mars 1964 - 1971

Early in 1959, Pioneer 4 discovered the Earth's planetary magnetic field in space. Later that year, Luna l, 2 and 3 (Russian) and Pioneer 4 all made flybys orbiting the Moon. Vanguard 3 studied solar radiation in space. An exciting era for astronomy had just opened.

In 1960, Atlas 4 and 5 failed to reach the Moon, and Pioneer 5 sent data back of the solar wind in space. In 1961, the Soviet Venera 1 came within 6,000,000 miles of Venus. In early l962, Ranger 3 and 5 missed the Moon, but Ranger 4 landed on the Moon, but its cameras failed. The USSR's Cosmos 3 and 7 studied solar flares. Mariner 2 made a flyby of Venus at 20,000 miles. Mars 1 (USSR) missed Mars.

In 1963, Explorer 18 orbited the Earth at 150,000 miles, monitoring solar flares. Two more USSR space probes missed the Moon. In 1964, Ranger 6 landed on the Moon but again the systems failed. Next, Ranger 7 and 8 landed on the Moon and also sent over 11,000 photographs of its surface.

Russia's Zond 1 approached Venus. Zond 2, sent to Mars, failed in 1965. In November, 1964, Mariner 4 finally succeeded in making a close flyby of Mars and transmitted 21 pictures of its surface before contact was lost. More data was needed. As to the question of life on Mars, the results became increasingly unpromising.

In February 1969, Mariner 6 was launched and made a successful flyby of Mars, transmitting 75 pictures. It approached within 2,100 miles of its surface. A month later, Mariner 7 transmitted 126 pictures also from 2,100 miles of the Martian surface. The technology was exciting but it was a bad year for those yearning for life to be found on Mars. Later that year, Armstrong walked on the Moon and brought back some samples of Moon rocks.

1971. Two years later, Mariner 9 flew within 2,100 miles of Mars and transmitted returned 7,326 images of the surface of Mars. These images revealed unimaginable scars of catastrophism, indicating a surface of Mars tortured by craters, racked by rifts, battered with bulges, tried by extreme temperatures and with wild volcanism gone amok.

And further, mysteriously, most surprising all, its two tiny satellites, Deimos and Phobos, were badly poxed with pitlets. As was mentioned earlier, there was inadequate theory as to why Mars had these two satellites. But for their both being poxed with tiny pitlets, there was no theory. All this was unforeseen. For those yearning for life on Mars, it was a horrible year. The legend was over. Almost.

In 1976, Viking 1 and 2 landed on Mars, and returned more data. Its data underscored how inhospitable, cold and dry is the surface of Mars with respect to water-based life, as it is known on our planet.

Nevertheless, the latest page in the story of the surface of Mars is not yet written. In 1996 or 1997, Mars Pathfinder will enter the red planet's thin atmosphere, and parachute an instrument package toward the surface. Close to the ground, air bags will inflate around the package, insuring a soft landing.

What is the purported purpose of Mars Pathfinder - the big reason for its funding? It is to lay the groundwork for future efforts to find FOSSIL REMAINS OF LIVING THINGS.

Scientists in 1996 are somewhat shifty on the topic of life on Mars because there is so much wishful thinking, but so little (no) evidence. But there are scientists who want to promote space exploration and badly need a reason for massive funding for this and other projects. Space mission research is an industry, and it needs a flow of projects, and reasons for more funding. But alleging there are fossils of former life on such an inhospitable planet is troubling.

Well, yes, and most scientists today believe that Mars is barren of life. But that doesn't mean that there was never life there. And if life was there once, there's always a chance that we'll be able to find traces of it. [n6]

First of all, astronomers need to understand planetary catastrophism. The theory of Astra's fragmentation, 75,000,000 miles from the present orbit of Mars, also predicts that there is some residual debris still in orbit near its Roche Limit. This is 2,500 miles and more above its surface, and on a plane that is an extension of its equator. With poor routing, a multi-billion dollar Mars mission could encounter that expected debris and be destroyed.

Second, NASA officials fail to understand how recently were the Mars-Earth Wars, and before that, the delivery of Mars and other planets to the realm of the Sun. They fail to understand the evidence as to how recently the Sun experienced a nova, or why. Mars missions should be funded, but only when sound reasons are forthcoming.

Investigating paleomagnetic polarity reversals on lava flows in Mars could be a reason Mars Pathfinder should be funded. As was mentioned earlier, another reason is to analyze ices yet remaining in the subsoil's, in places in the Eastern Hemisphere of Mars. The Martian ices need to be assayed to determine if they also lack the oxygen 18 isotope, and the deuterium form of hydrogen.

In 1996 AD, ice still exists, 3,500 feet thick, all below modern sea level, resting on Antarctic bedrock. It is intermixed with volcanic ash, has a uniform crystalline axis, and is a strange, unearthly type of ice. Its oxygen lacks the oxygen 18 isotope. Its hydrogen contains no deuterium, much different from water in the Earth's oceans and snows in its polar regions. Perhaps a similarity will be found between Mars ice and the deep sub-sea level Antarctic ice. Perhaps such an investigation would help groping geologists form a clear, consistent ice age theory.

Astronomers and geologists wonder how Mars could have possibly had a sudden, widespread flood if there is no oxygen, no water vapor and only a tiny bit of nitrogen there. Isn't a planetary flood a kind of topic, a tradition, or perhaps even a legend or myth that applies only to the Earth? And hasn't that flood tradition, involving Noah, or Utnapishtim, however widespread, been widely discredited as "impossible" according to gain sayers in gradualist geology?

Evidence Of The Flood Of Mars

NO IRRIGATION DISTRICTS. In February, 1969, Mariner 6 made a flyby of Mars at 21,000 feet and transmitted 75 pictures of its surface. It was followed by Mariner 7 a month later, which sent 126 pictures. Later, in 1971, Mariner 9 sent over 7,300 pictures of the surface of Mars. No canals were found. No running rivers. No reservoirs. No flumes or siphons. No oases. No irrigation district headquarters. No water ice. No liquid water.

MISDIRECTED PUBLICITY AND CLAMOR. It turns out that for the first 60 years of this century, those scientists who viewed Mars and didn't see lines, or canals, or oases, had been the less vocal. Or perhaps they just got very little press. But they were the ones to whom the ears in astronomy and journalism should have been attuned.

At the surface of Mars, gravitational attraction is only 30% of what it is at the Earth's surface. On Mars, an earthling might high jump 20 feet high, and a pole-vaulter 50 feet. Now dry ancient river beds are in evidence across the cold surface of only one hemisphere of Mars.

> Drainage networks. Dense, dendritic networks of channels such as this are a common feature of the southern highlands of Mars. Dendritic patterns do suggest a fluvial origin of the channels ... [n7]
>
> It is possible to recognize three kinds of channel: (sic) run-off channels, outflow channels and fretted channels. Run-off channels typically have a V-shaped cross section, start small and increase in size downstream, and have well-developed tributary networks. The large branching channels Nirgal and Ma'adam Vallis can be assigned to this group, but more typical are large numbers of smaller networks that incise the ancient cratered terrain. [n8]

POST ASTRA IN TIMING. The timing of the Astra fragmentation has not been tied down, yet prospects are good that it can and will be done. Whatever the timing of Astra's demise, the icy fragmentation causing these flash rivers was later.

The later dating is established because a dry river bed is found where its ancient river flowed into one side of an older crater, filled it up, and then flowed out the other side again, same flow volume.

TWO ICY FRAGMENTATION'S. If the hemispheric geography of the dry river beds is correctly understood, and if the suddenness in the appearance and outflow of those rivers is correctly understood, the Eastern Hemisphere of Mars suffered a sudden icy spray from space. Recently, in terms of astronomical time, that is.

This means our Solar System has experienced not just one, but two icy fragmentation's. One on the Roche Limit of Saturn, and one near Mars. One resulted in Saturn's icy rings, which are at its Roche Limit. The other produced the dry river beds of Mars.

This duet of icy fragmentation's accompanies a quartet of rocky fragmentation's, the dark ring systems of Jupiter, of Uranus, of Neptune and the former rocky ring system of Mars. Suddenly it appears that this Solar System has a history of no less than six "little bangs".

Two of the Little Bangs, one icy in nature and one rocky, involved our close neighbor, Mars. Astronomers may or many not be correct about the Big Bang theory. But it is sure that they have missed both Little Bangs involving our close neighbor, Mars.

FLOWING WATER VELOCITIES. Flowing water velocities of 20, 30 and 40 mph are in evidence.

A FLASH FLOOD. Sudden catastrophic flooding is in evidence. One outflow channel, Capri Chasma, is compared in its dimensions to an ancient flood across Eastern Washington.

That prehistoric outflow occurred when an ice lobe loped out of mountain valleys of Southern British Columbia, and into Northeastern Washington. In so doing, the large ice lobe dammed up the combined drainage's of the Columbia River, the Kootenai, the Clark's Fork, the Flathead, the Coeur d'Alene, the St. Joe and the Spokane Rivers.

The intermontane ice lobe that galloped out of a mountain valley from British Columbia dammed all drainage between Trail B.C. and Spokane. Water was backed up some 250 miles, to within the shadows of the Continental Divide. A lake with a surface area of nearly 20,000 sq. miles formed. Its surface was nearly as big as Lake Michigan, another ice age lake.

When the ice began melting, the melting started from the bottom, where the ground had been warm. When the ice dam broke, it was sudden. Thousands of cubic miles of water gushed out and rampaged southwesterly across what are now called the scab lands of Eastern Washington. They galloped at velocities of 60 mph and higher.

The prehistoric great glacial lake, known as "Lake Missoula", drained in 48 hours. Sediments, gravel, rocks and even boulders in this rampage were entrained and carried downstream. Some of the boulders silted out as far away as 250 miles, near Portland. Larger stones silt out first, as water velocities begin to ebb. Lake Missoula was transformed into an "outflow channel." It also contributed to the silting of the Willamette Valley on Western Oregon.

Like Lake Missoula, Capri Chasma on Mars also was an "outflow channel." It was formed when a flash flood filled a huge lake, whose banks suddenly broke. Chasma's waters rampaged 50 and 60 mph across the adjoining plain. Outflow waters spread out, as waters tend to do when a dam breaks.

The width of Capri Chasma is defined as where it broke, and swept aside its dams walls. It was 35 miles wide. On the Earth, the widest river is the Amazon, 70 miles wide at its mouth. It is rain fed and drains a humid tropical region of 2,000,000 sq. miles. In comparison, at its widest, Capri Chasma was 35 miles wide. Capri Chasma was, half the width of the Amazon, and wider than the mouth of any other river on the Earth. This was "flash flood catastrophism."

Another type of dry river bed is a "run-off channel." Run-off channels gather tributaries in their upper course and widen downstream, like the Mississippi River or the Danube. Most rivers on Earth are rain and snow fed, annually, and are of this type. Their currents are usually under 5 mph.

Ma'adam Vallis is also a "run-off channel," about 400 miles long. Its water velocities averaged 25 mph. This velocity was achieved on a planet where the surface gravitational force is but a third of that force at the Earth's surface.

Being 400 miles long, with velocities of 25+ mph, it can be calculated that Ma'adam Vallis rampaged for some 16+ hours or so. It is one of the longest dry river beds in the Eastern Hemisphere of Mars. It rampaged across cold surfaces with deep sub zero temperatures, and apparently its waters, originally hot, cooled and shortly froze. With deep subzero night time temperatures, and a cold surface, those waters, originally hot, seemingly froze their first night on Mars.

Night time temperatures on Mars frequently are around -190° F. If ice fragments sprayed Mars, the energy of motion of the ice fragments would convert to heat the instant of crustal impact. Those waters then cooled from near boiling on the surface of Mars to freezing in less than a day. Like Chasma, Ma'adam Valles also was something less than a 24-hour affair.

A third type of dry river bed is the "fretted channel," with a flat floor, steep walls, and frequent changes in direction. One such fretted channel, is in the Ismenius Lacus region of Mars. Details of Ismenius and the terrain it crosses, flowing into a crater and out the other end, same volume, 120 miles long, again indicates a flash flood.

As a collective show, the three or four dozen dry river beds of Mars indicate that the Flood of Mars was a sudden, sub-24-hour affair. Their genesis was sudden and their freezing was quick. In addition, the flood was spotty, and it was hemispheric in scope, not global. The spotty hemispheric scope of this flash flood reminds one of the spotty geographical scope of the fragments of Astra, which also hit the Martian surface on one side only.

On the surface of Mars are found, alas, no canals, but it has ample indications of a hemisphere-sized flash flood. To repeat, the fact that no river beds are longer than 400 miles, coupled with the fact that they rampaged at high velocities, indicate that these waters ran across the cold surface of the red planet until, the first night with surface temperatures below -150° F., they froze.

Percival Lowell predicted canals with slow-moving water, flumes, gauges, and reservoirs (or oases.) It would have been better had some 19th century newspapers reported catastrophists such as Louis Agassiz on the ice age. Or on Baron Cuvier and the sudden sedimentation of the Paris Basin. Or on George McCready Price and the velocities of water needed to entrain various sized erratic boulders up to 150 tons.

The evidence of flash flooding is there, but what was the sudden cause? It was a sudden icy spray from space, by a fragmenting ice ball.

The Fragmentation Of An Ice Ball

The Second Of Two Fragmentation's Of Scar-Faced Mars

Old Scar face experienced a rocky fragmentation, that of Astra. It was of the type of the fragmentation that produced rocky debris. Other fragmentations at the Roche Limits of Jupiter, Uranus and Neptune produced permanent dark rings. But Saturn has an icy ring system, indicating its fragmentation was by a cold, straying ice ball.

THE SOURCE. What fed the rampaging rivers of Mars? The source seems to have been external; Mars cannot retain water vapor due to molecular escape velocities and due to its small mass. The most likely answer is similar to what happened to Saturn. It was another fragmentation of an ice ball on another Roche Limit, that of the red planet.

One difference is that Saturn has 885 times as much mass as Mars. A second difference is that Saturn is more than four times as distant from the Sun as is Mars. A third difference is that Saturn's perihelion is, and was 830,000,000 miles from the Sun whereas, in this model, the Catastrophic Third Orbit of Mars had its perihelion at a toasty 64,350,000 miles. See Tables XI, XII and XIII.

The evidence is that an ice ball fragmented above the Eastern Hemisphere of Mars. Its icy fragments sprayed a large part of one entire hemisphere of Mars. Upon impact, the energy of its motion converted instantly to heat. It vaporized and condensed.

A flash flood came to the Eastern Hemisphere of Mars. (1) Its hemispheric scope suggests this. (2) The scattered, or spotty distribution points to this. (3) The velocity of the flowing water indicates suddenness. (4) The minuscule atmosphere of Mars also points to space as the region of origin for its former ices.

THE ICE BALL'S NAME. Astronomy traditionally accords to the discoverer the privilege of naming a new satellite, planet or star. Our nomination of this second ice ball sprayed Mars with a spray of ice, is "Glacis". Glacis in French means "ice."

Part of the evidence concerns the ice fragments that hit little Mars. If their fragment velocities were in the tens of thousands of mph, there was very little atmosphere to create any friction during their brief descent. Such is in contrast to the atmosphere of the planet Earth, and its atmosphere, 100 times thicker. Due to friction created by our atmosphere, ice fragments will "burn". A similar spray for the Earth would produce a sudden, warm rain across one entire hemisphere. It would also produce explosive fragmentations such as the Tunguska bolide, June 30, 1908.

For Mars, technically it cannot be said that ice can burn. Ice fragments can suddenly hit its surface and come to an instant halt, convert the energy of motion into heat, which causes sudden vaporization, and next a sudden, massive condensation. To the extent water can boil on Mars, this ice vaporized, condensed and began to rampage to lower elevations with velocities similar to a suddenly wounded, wild water buffalo. Such is the direction in which the evidence points.

Gradualists On The Dry River Beds Of Mars

TIMING AND CAUSE. It is of interest to evaluate to what kind of a timing gradualists attribute to the rampaging rivers of Mars. It also is interesting to note their idea of the cause.

> North of the canyons are numerous fascinating outflow channels which appear to have been produced during a period of catastrophic flooding 3,000 to 3,500 million years ago. [n9]

Gradualists assume 3 billion years ago, give or take a half billion years. Imprecision.

Planetary catastrophists recognize that this was the second and the last occasion for a fragmentation to bother little Mars. It was recent, although exactly how recent has yet to be established.

If the Greeks are to be understood correctly in that their ancestors saw Astra fragment, from a vantage point afar off, then the dating of the demise of Glacis is within the last 10,000 years, conservatively. Moreover, if their earlier ancestors saw Astra fragment, their later ancestors quite probably saw both Mars approach and with it the Glacis ice ball approach. Maybe it was together. In the ancient account, it was frightening to say the least.

More evidence in projected Volume IV will discuss the flash flood of Mars in the light of our planet's greatest flood. Thus both planets have had floods, hemispheric in scope. Both were followed by ice ages, as is in evidence shortly. That evidence will produce a better understanding of ice ages of both planets.

In general, planetary catastrophists require about one five hundred thousandth of the amount of time to lapse for the ancient icy spray of Mars as do the gradualists. Seldom if ever contradicted, they assert with some confidence, three billion years, plus or minus a half billion.

THREE ISSUES. One issue is WHEN Glacis fragmented. A second issue is WHY. A third issue is WITH WHAT OTHER EFFECTS.

> The rapid release of water that could have produced such a flood channel MAY have resulted from melting of ground ice by volcanic activity beneath the Martian surface, or it MAY have issued from previously buried aquifers, although why this should happen has not yet been adequately explained. [n10] (Our caps.)

There is a troubling geographical question here that needs to be asked; and it is a simple question. 99% of the volcanism on Mars happens to have occurred in the Western Hemisphere of Mars, in the Tharsis Bulge region. 99% of the flash floods and the dry river beds of Mars are in its Eastern Hemisphere, with just one dry river bed being on the eastern edge of its Western Hemisphere.

On the surface of Mars, its dry river beds are 3,000 miles distant from the region of giant volcanoes, Olympus Mons, Arsia Mons, et al. Are gradualists trying to offer up to sensible people that volcanic flows, 3,000 miles distant, were the cause for once rampaging, now dry river beds? Apparently. And are they saying the source was subcrustal ices, 3,000 miles distant? And are they also saying "Suddenly" for this? Any inspection of such an explanation is troubling.

Mars has 98% or 99% of its dry river beds in one hemisphere, and only 2% of its volcanoes. And vice versa. What kind of a geographical explanation is that? What kind of a scientific explanation is it? Gradualists are advised to rethink this entire issue and come up with something more realistic - or do the unthinkable alternative, become planetary catastrophists.

Short Term Icy Comets

Since the Eastern Hemisphere of Mars was sprayed by icy fragments of Glacis from space, it is likely that most of the spray missed little Mars. If so, the ice fragment would have proceeded out into space --- and the region of space which the old Mars orbited. What became of those icy fragments that missed Mars?

SHORT TERM ICY COMETS. Short term icy comets are tiny ice balls, usually dirty ice balls. If they come close enough to the Sun, solar radiation RAPIDLY effervesces away their ices into long streams of glistening water vapors - cometary tails.

Their vapors, long cometary tails, are then bent into curved streamer shapes as they are left behind, only to be blown by the solar wind. Thus, a comet's tail is formed. Normally, one side of the tail of a comet mirrors the other side, as it is with bird wings. There may be multiple mirrored streamers, or wings in a cometary tail. The cometary ices trail, and to point back to the comet's tiny, icy nucleus.

Those icy comets that are nearer to the Sun effervesce away faster, and have higher attrition rates. Comets farther out survive longer. How many icy fragments have effervesced away in the last 5,000 years? Nobody knows. But it was many, based on modern comet attrition rates.

Even the last 150 years has demonstrated a high mortality rate among the short term icy comets. Some are disappearing each decade. The count of the short term icy comets, entirely within Jupiter's orbit, now number only about 100. [n11]

The genesis of the short term icy comets is disputed.

> According to a theory developed by J. H. Oort in 1950, there is a diffuse cloud or reservoir of gas, dust and comets that is gravitationally part of the Solar System but about 40,000 a.u. distant from the Sun. ...
>
> There are some astronomers, notably V. Clube and W. Napier of the Royal Observatory, Edinburgh, who believe comets to be of interstellar origin, so they are captured by the Sun instead of being original members of the Solar System. This remains at present a minority view, but cannot be discounted; there are many questions about the origin of comets that remain to be answered. [n12]

> Short-period comets have characteristic lifetimes of between a few hundred and a few thousand years. Not only do they break up, they also get driven away by planetary encounters. There are at present approximately one hundred times too many short-period comets relative to the rate at which long-period comets are captured by Jupiter and fed into the observed stock of Apollo asteroids. The present number is probably due to the burst of new short-period comets formed several thousand years ago as a result of a single large comet fragmenting during Jovian capture or perihelion passage. [n13]

Contrary to the cited explanation, the fragmentation of Glacis and the spray of icy fragments was not due to Jupiter. Like the demise of Astra, it was due to the fragmenting of Glacis on the Roche Limit of an inner planet. Glacis fragmented on the Roche Limit of either the Earth, Mars or Venus.

A Third Theory

A pair of theories on the genesis of the short term icy comets was cited. To this pair now a third theory is added. It is more likely that the genesis of short term icy comets are, in a sense, second generation. First an ice ball was imported into the inner region of the Solar system, and later it fragmented. This analysis is that initially, the co-orbiting Mars-Glacis was delivered to the realm of the Sun by Little Brother (see Vol. I).

Next, the short term icy comets have been simply survivors of an ancient icy fragmentation of Glacis on the Earth's Roche Limit during the closest of all Mars flybys. The genesis of the short term icy comets is that they are icy fragments that (1) missed Mars, (2) missed the Earth, and (3) have had enough icy mass to not yet effervesce away in the last 4,500 years. No doubt the majority already effervesced away. The surviving icy comets acquired new orbits, but from the energy of the old orbit of Mars.

Modeling The Fragmentation Of Glacis

When Glacis fragmented, it sprayed the Eastern Hemisphere of Mars. Mars was in its old catastrophic orbit of the earlier age. The Roche Limit of three planets are the three candidates, of (1) Venus, (2) the Earth and (3) Mars. Of these, if two can be eliminated, then the remaining Roche Limit is the cause of the fragmenting of little Glacis.

THE ROCHE LIMIT OF VENUS. A Mars-Venus flyby, with Mars towing the ice ball, is theoretically possible. But the orbit of Venus is too close to the Sun for the remaining short term icy comets. Venus in that era was 67,500,000 miles from the Sun. Very few cometary orbits come close to Venus.

THE ROCHE LIMIT OF MARS. If Mars was towing Glacis, Glacis would have had an orbit close in, something like Phobos at 5,700 miles. Otherwise it would have been swept out and away from Mars by either the Sun, or by flybys of Venus and/or the Earth.

Were Glacis to come within 5,000 miles of Mars, as did Astra, it would produce an icy demise, like Astra. Orbiting Mars and penetrating its Roche Limit are two different conditions. Deimos comes within about 15,000 miles of Mars, and it appears that Mars almost lost it. If Mars was co-orbiting with Glacis, and was in a sense towing Glacis, the Mars Roche Limit probably is not the cause for the shattering of Glacis.

THE ROCHE LIMIT OF THE EARTH. Most of the evidence of which we are aware points to the Earth's Roche Limit as the cause for the shattering of Glacis. It was during the closest of all Mars flybys, estimated at 15,000 miles. A foundation for this distance will occur in Volume III.

This opens up the possibility that the shattering of Glacis sprayed the Eastern Hemispheres of both the Earth and Mars simultaneously. This choice is attractive because it agrees with the bulk of the evidence of which we are aware, and it begins to answer (a) the sudden rain on the opening day of Noah's Flood, plus (b) ice age theory for the Earth.

The evidence for the third candidate, the Earth's Roche Limit, comes partly from space, the short term comets, and the dry river beds of Mars, and partly from the Earth's crust, from Antarctica, Siberia, and Alaska, and partly from ancient literature.

Some of the evidence from ancient literature comes from Hebrew Talmudic sources. Those literary resources were gathered by studious Jews while during their Babylonian exile. It was circa 550 B.C.E., almost 2,000 years after Noah's Flood.

Those Neo-Babylonian sources reflected both earlier Chaldean sources and earlier Sumerian sources, of which original writing, or copies thereof, was still available on cuneiform tablets. That material relates to the topic of the cause of Noah's Flood.

> The flood was produced by a UNION of the male waters, which are above the firmament, and the female waters issuing from the earth [OCEANS]. The upper waters rushed through the space left when God removed TWO STARS out of the constellation Pleiades. Afterward, to put a stop to the flood, God had to transfer TWO STARS from the constellation of the Bear to the constellation of the Pleiades....
>
> There were other changes among the celestial spheres during the year of the flood. [n14] (Caps ours).

One of those other celestial changes was a shift in the location of the spin axis, a precession.

Chaldean, Sumerian and Assyrian literature on Noah's Flood was still on clay tablets, probably in copies from the original accounts, second generation copies. In the 19th century and early 20th century AD nine flood stories were discovered in cuneiform on clay tablets. These included from a massive library of 20,000 tables, Assurbanipal's library.

Three flood accounts are in Assyrian cuneiform, three are in Chaldean and three are in the oldest language, Sumerian cuneiform. Some accounts are incomplete because of damage to the tablets. The Epic of Gilgamesh is the most complete of those ancient cuneiform accounts of Noah's (Utnapishtim's) Flood.

TWO STARS, NOT ONE. Ginzberg's sources affirm that two stars, not one, approached the Earth on the day of the onset of Noah's Flood. This model states that Mars was one. So does the Epic of Gilgamesh. The year was 2484 B.C.E. It was on target for the 108-year cyclicism. Evidence of the 108-year cyclicism is in Volume III.

There is other evidence one of the stars was Mars. Sumerian commentary from The Epic of Gilgamesh expressly identifies Enlil (Mars) as THE CAUSE of Noah's Flood. Gilgamesh's eleventh tablet is an extensive description of the onset of Utnapishtim's (Noah's) Flood. It is four times as long as is the Genesis account of Noah's Flood.

> (But) Enlil shall not come near to the offering,
> Because without reflection HE BROUGHT ON THE DELUGE
> And consigned my people to destruction!
> As soon as Enlil arrived and saw the ship, Enlil was wroth;
> He was filled with anger against the gods, the Igigi:
> "Has any of the mortals escaped? No man was to live though the destruction!" [n15]

In the Sumerian pantheon, Enlil was Mars. Ea was the Earth. Innanna was Venus. Anu was Jupiter. Ninurta was Saturn.

What was the other star that also approached the Earth? Was it Glacis, an icy satellite with a diameter of 500 to 600 miles? Was it shattered ice, coming in from a fragmented Glacis, that entered our atmosphere, burned, and recondensed into a hemisphere-wide warm rain? Talmudic evidence points in this direction.

CLIMATOLOGY. The Inuit, Eskimos of Alaska, remember in their lore a time when volcanism suddenly erupted, and the Sun went low in the sky. This suggests a radical shift in latitude. Before that, it was easy to make a living, but afterward, it was only with difficulty. This makes sense if there was a sudden torque on the Earth's spin axis, and a radical shift in latitude resulted.

Hebrew Talmudic literature is as follows:

> According to BR 32. 11, and the parallels cited by Theodor, Noah suffered very much from the cold, while the same source, 9, and Tehillim 1, 11, state that the ark, despite the mass of water, rested quietly "as a ship in port". ... Zohar I, 68a-68b and 69a, states that Noah by hiding in the ark escaped the angel of death. [n16]

If this third candidate, the Earth's Roche Limit, is the best answer, it follows that the following sextet of topics as best being understood in tandem, and as being simultaneous.

1. The genesis of the dry river beds of Mars.

2. The genesis of the sudden rain in the Eastern Hemisphere of the Earth during Noah's Flood.

3. The genesis of the immense tides emerging out of the Indian Ocean during Noah's flood, floating the Ark but causing vast damage to Southern Asia.

4. The genesis of the Earth's last ice age, and perhaps its only ice age.

5. The genesis of the short term icy comets, icy fragments that missed both Ares and Hera, Mars and the Earth.

6. The genesis of ice found yet today 3,000 to 4,000 thick below sea level, resting on Antarctic bedrock. This ice is deficient in the oxygen 18 isotope common to ocean water and clouds. Also the ice is deficient in the deuterium form of hydrogen.

   It has a uniform crystalline axis, pointing about 10° from vertical, suggesting it was related to, or directed in by magnetic field force lines. It is also intermixed in the volcanic ash, 3,000 and 4,000 feet below modern sea level, indicating that the polar flat spot was not at Antarctica at the time of Noah's Flood. As to volcanic ash, the deeper the ice core drilling, the thicker the ash in the mixture.

If this answer is the correct answer, ice in the nuclei and in the tails of short term icy comets should be found also with deficiencies in the oxygen 18 isotope and deuterium just like Antarctic sub-sea level ice. In addition, if sub-surface ice on Mars is found and assayed, it also will contain these deficiencies.

If this is the correct answer, this also opens up the possibility that the ice, descending in vast volumes over the magnetic poles, was at temperatures close to -300° F. It would be well to take another core drilling in deep sub-sea level Antarctic ice to have its temperature taken. Some of it might still be surprisingly, even unearthly cold, like the ices of Saturn's Rings, Callisto, Ganymede, etc.

This chapter has introduced a new theory on the genesis of short term icy comets. It has introduced evidence of flash floods, Lake Missoula and Capri Chasma. The subject of the ice ages of Mars and the Earth is one of vast length, breadth and depth.

In Volume IV, a foundation, begun to be laid here, is hoped to be completed in three years. The subject of ice age genesis is sufficiently massive and interrelated to require several volumes, partly involving geology, partly involving oceanography, and partly involving astronomy.

The first conclusion is that this sextet of "genesis issues" all originated and began to unfold on one day in late October in the year 2484 B.C.E. The day of that month was the 24th on modern calendars. It was the 17th of Marchesvan, "Mars' month", on the ancient Chaldean calendar, the Tishri calendar.

Marchesvan is cognate with the Hebrew word for Mars, "'ma 'owr", and with the Roman "Mars" and with the Greek "magna Ares." On the old Tishri calendar, the third month was Kislev, named after Jupiter. The ancient Tishri calendar began on the new moon of September 7.

The second conclusion is that on the day of Noah's Flood, it featured a strange union of waters from below, which were monster tides from the Indian Ocean, and an abundance of warm waters from above. They were warmed by friction in passing through the Earth's atmosphere, originally fragments from Glacis.

The "waters from below" were massive tides erupting out of an Indian Ocean. It contains 28,000,000 cubic miles of water, about 25% of the water on the Earth's surface. In the Book of Genesis the flood came from "the fountains of the great deep." Waters from the Indian Ocean and from the "fountains of the great deep" were the same thing.

With 500 to 600 miles as an estimate for the diameter of Glacis, it was similar in size to that of Saturn's Enceladus (320 miles) or Tethys (650 miles). It also was much like the Uranian Umbriel (500 miles) or Ariel (650 miles). And its ices may have been almost as cold.

The third conclusion is that indeed on this particular flyby, the Earth was assaulted by not one "star" [planet] but a planet and a satellite [Mars and Glacis.] Talmudic commentary is very clear on this count, whether correct or otherwise.

This particular flyby in October, 2484 B.C.E., was the closest of the long series of Mars flybys, for two precise reasons. Those reasons are identified and are discussed in two chapters in the projected Volume III, entitled "The Flood of Noah". Meanwhile, the advice needed for groping gradualists is that one can delve deeper into Genesis, Job, Isaiah and Talmudic commentary with profit.

> The view that the flood was brought about by a union of the male and female waters is found also in Enoch 54. 8-9, and goes back, as has already been remarked by others, to the Babylonian conception of Asp and Teammate. Comp. Yerushalmi Berakot 9, 14a, Yerushalmi Ta'anit 1, 64b, Babli 6b; 6b Tosefta 1.4; Tehillim 42, 267; PRE 5; Koheleth Z., 87. [n17]

Obviously, Hebrew Talmudic commentary about conditions and events surrounding Noah's Flood came from much earlier sources. It was from cuneiform tablets written in Assyrian, Babylonian, Chaldean, Sumerian and possibly Persian sources.

On the surface of the Eastern Hemisphere of Mars was added much surface ice by the fragmentation of Glacis. Those waters collected in a spotty geographical pattern of river beds and lakes to be sure. But for one day, and only one day, the rivers appeared, rampaged roughshod, and then promptly froze. In their place on the surface of the Martian Eastern Hemisphere was formed frozen lakes, frozen ponds and frozen river beds.

Where did those ices dumped on the Earth go? In the Earth's Ice Age, the ices descended from space and formed huge dumps, dumps close to the then magnetic poles. From the Keewatin Node in Upper Manitoba outflowed a lobe so great it deposited vast volumes of fine Canadian soil across the American Midwest from Montana to Ohio.

The edges of the most advanced lobe formed the Missouri River Valley and the Ohio River Valley. It formed Lake Superior, Lake Michigan, Lake Huron, Lake Erie, Lake Ontario and one lobe reached as far south as Cape Girardeau Missouri, at latitude 37. There, the end of the ice lobe was well over half the way from our North Pole to the equator.

Another flow came down the Frazier River Valley. It was a crunching flow, 2,000 feet deep on the Cascade Mountains east of Seattle, according to glacial striations.

Yet a third lobe formed on the surface of Labrador and Upper Quebec, the Labrador node. It flowed across Quebec and across Upper New York State. This lobe flowed not only around the Adirondack Mountains but also OVER THEM. They are 2,000 and 3,000 feet above sea level. Terminal moraines from the Labrador node were massive enough to form the Cape Cod peninsula and Long Island.

Other ice masses formed in Northern Europe and on the surface of Antarctica, then above sea level but now 4,000 feet below modern sea level. New Zealand was dealt ice flows, but Alaska and Siberia, farther from the magnetic pole, largely escaped.

This model indicates that when these ices finally melted, the Earth's oceans increased in volume about 7%, and simultaneously the mean temperature of the oceans decreased some 10° F. The Earth's last ice age was a big, cool case of catastrophism from the celestial region, a region where more ice exists, and it is both cold and abundant.

Ices effervesced off Glacis before 2484 B.C.E. and then off the surface of the Eastern Hemisphere of Mars after 2484 B.C.E. Other ices effervesced from the nuclei of the short term icy comets, some of it continuing to this time.

The Golden Fleece Of Aries

The surface ices on Mars had nowhere else to go other than to effervesce and float off into space during the Martian orbits past perihelion. Then, the distance at perihelion was a toasty 63,350,000 miles from the Sun. As a result, there was a shift in location of the Martian ices. They left the surface of Mars' Eastern Hemisphere effervesce and floated off into space. The effervescing ices from the surface of Mars was one of the Scars of Mars. Such is the origin of what the Greeks called, "The Golden Fleece of Aries."

Sometimes, Mars was 228,805,000 miles from Sun, at its ancient aphelion. Temperatures were such that little if any of its ices effervesced there. But when ancient Mars approached to 90,000,000 miles from the Sun, its visible cometary tail began to grow, and began to spread out, trailing Mars in its ancient orbit.

As Mars came in to 70,000,000 miles, solar radiation increased, and the cometary tail of Mars expanded in length and breadth. The Earth's sunward side might well have been swept lightly by the icy gauze from Mars. Figure 14 illustrates.

At 64,350,000 miles from the Sun, slightly inside the orbit of Venus, the rate of the effervescing of the ices doubled and redoubled, increasing rapidly, if briefly. If Venus happened to be nearby, its face also might have been swept and rinsed by the icy gauze. This was near the Mars perihelion, where maximum effervescing occurred.

But it was not where the maximum cometary tail of Mars fully developed. That was weeks later as Mars was approaching its March 20-21 flyby site of the Earth. A foundation will be laid in Volume III that once every 108 years, almost like clockwork, the Earth was close at hand. The Jews called it "Passover", the time of the destructive angel's Passover. The Romans called it "Tubulustrium," the time of trouble.

Figure 14 - The Golden Fleece Of Aries - Grand Scale

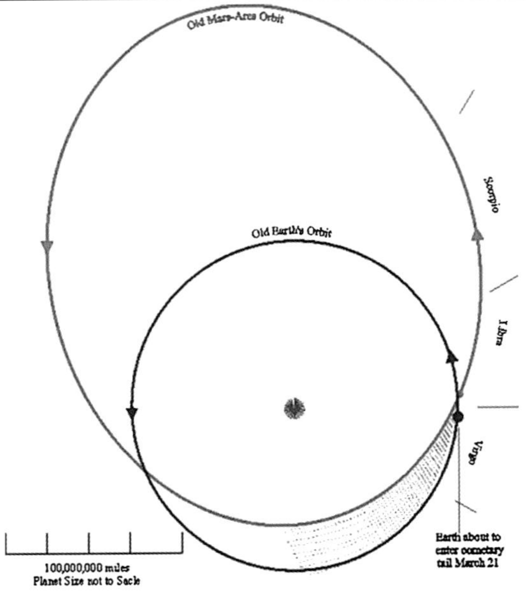

The Golden Fleece of Aries
The 100,000,000 Mile Plus Long Cometary Tail of Mars
During a March 21 Flyby

In this March crossing geometry, celestial mechanics and resonance orbit principles require that Mars always passed by the Earth on the sunward side, except the Final Flyby. As Mars did so, its cometary tail swept across the face of our planet. Visually, this is part of the reason that made the ancient Passover scenes and tubulustrium scenes so spectacular, so memorable, so frightful.

The ancient, magnificent cometary tail of Mars was called by the Greeks "The Golden Fleece of Aries" and it was for several reasons. First, Ares was Mars. Second its cometary tail was in full bloom as it swept across the face of the Earth, in the month of Ares.

Third, "Aries" or Ares was entering the first zone of the zodiac; it was a 30-degree zone also called "Aries." Fourth, the vernal equinox and the first day of spring always occurred on March 21. On this day, the first star on the horizon at dawn appeared a small star, Mesartim, known to ancients as "the First Point of Aries." Mesartim is the lead star of the small, four-star constellation of Aries. Mesartim's appearance on the horizon at dawn heralded a new spring and a new year.

Fifth, this was the location in space where the Earth's sunward face could be swept by the fullness of the effervescing cometary tail of Mars (Ares.)

In that age, Mars had a 720-day orbit and it made a biennial pass across the Earth's orbit whether or not the Earth happened to be there. Once every 108 years on schedule, about 1% of the time, the Earth was close by when Mars also passed over.

In the Sumerian map of the heavens, the first 30-degree sector in space, their zodiac, was Aries. Each sector had a symbol. Usually symbols were animals but sometimes the sign of the sector was a pair of twins, a virgin, or a water bearer. The sector beginning on March 21, the first sector of the twelve, was symbolized by a ram with fighting horns, Aries by name.

Sheep skins in that era could be bleached into an off-white color. However, the "fleece of Aries" was not an off-white celestial fleece. In some accounts it was golden; in other accounts it was "the color of electrum", a pale yellow alloy of gold and silver. Probably the "Fleece of Aries" appeared more the color of electrum. There was the sunshine reflecting off it, and to some extent the Sun's rays shone through it.

It was this glistening gauze that trailed the chariot of Mars, wheels rotating, pulled by its [his] two trusty steeds, Phobos and Deimos. Those close Mars flybys were fearful, inflicting mass damage by fire, flood and earthquake, throwing out high voltage lightning toward the Earth, causing volcanoes to erupt, causing earthquakes, causing radical tides and tidal waves at the sea side.

Dreadful as the occasions were, destructive as they were, ugly as the face of Mars was, its long, glistening cometary tail threatened to sweep across the face of the Earth. To a neutral observer in space, it must have been a scintillating sight. To an Earth dweller, the sight was far from funny. Such was the terror of the ancient authors who saw the "fleece of Aries" and its associated nucleus, Mars.

> It was he, Herakles, who killed Kyknos, high-hearted son of Ares, for he came upon him in the precinct of Apollo, who strikes from afar, himself and his father, Ares insatiable in battle, BLAZING BOTH OF THEM LIKE THE LIGHT OF BURNING FIRE IN THEIR ARMOR, ... [n18]

The following citation describes more the visage of Ares than its cometary tail:

> In his hands he took up his shield, all-glancing, nor could anyone break it, either by cast or stroke, A WONDER TO LOOK AT. FOR ALL ABOUT THE CIRCLE OF IT, WITH ENAMEL AND WITH PALE IVORY AND WITH ELECTRUM IT SHONE, AND WITH GOLD GLOWING IT WAS BRIGHT, and there were folds of cobalt driven upon it. In the middle was a face of Panic ... [n19]
>
> ... and these were in silver, but the fir trees they had in their hands were golden,

> and they were streaming together, as if they were alive, and battering each other in close combat with spears and fir trunks. And on it were standing the swift-footed horses of grim-faced Ares, IN GOLD, and he himself, the spoiler, the destructive, ... [n20]

The Hebrew Angel Of Death And Its Wings

When the Eastern Hemisphere of Mars faced the Sun, near perihelion, ices effervesced abundantly for 10 to 12-hour periods. But when Mars turned its Western, or non-icy hemisphere to the Sun, no ices effervesced. Thus the effervescing of ices from Mars occurred in a daily pattern, in twelve-hour waves. To the ancients in Egypt and Mexico, it occurred that those waves of effervescing ices in the cometary tail of Mars resembled flapping wings, as in bird wings.

Thus it was that the Egyptians of that era had their Phoenix birds, which came close once or twice per century, threatening destruction, and flew with a feathery tail flying. The Mayans had their Quetzacoatl, also a fire bird. The Chinese had their fire star, Mars, and with its cometary tail, it was a writhing dragon's tail in appearance.

The cometary tail of Mars was seen and described by the Hebrews in yet another way. Today, angels and cherubs are considered sweet, friendly apparitions, but not then. Angels were messengers of the Lord, and one of them brought death and destruction in its train.

Figure 15 - The Passover Night Geometry I - (Also The Fleece Of Aries, Local Scale)

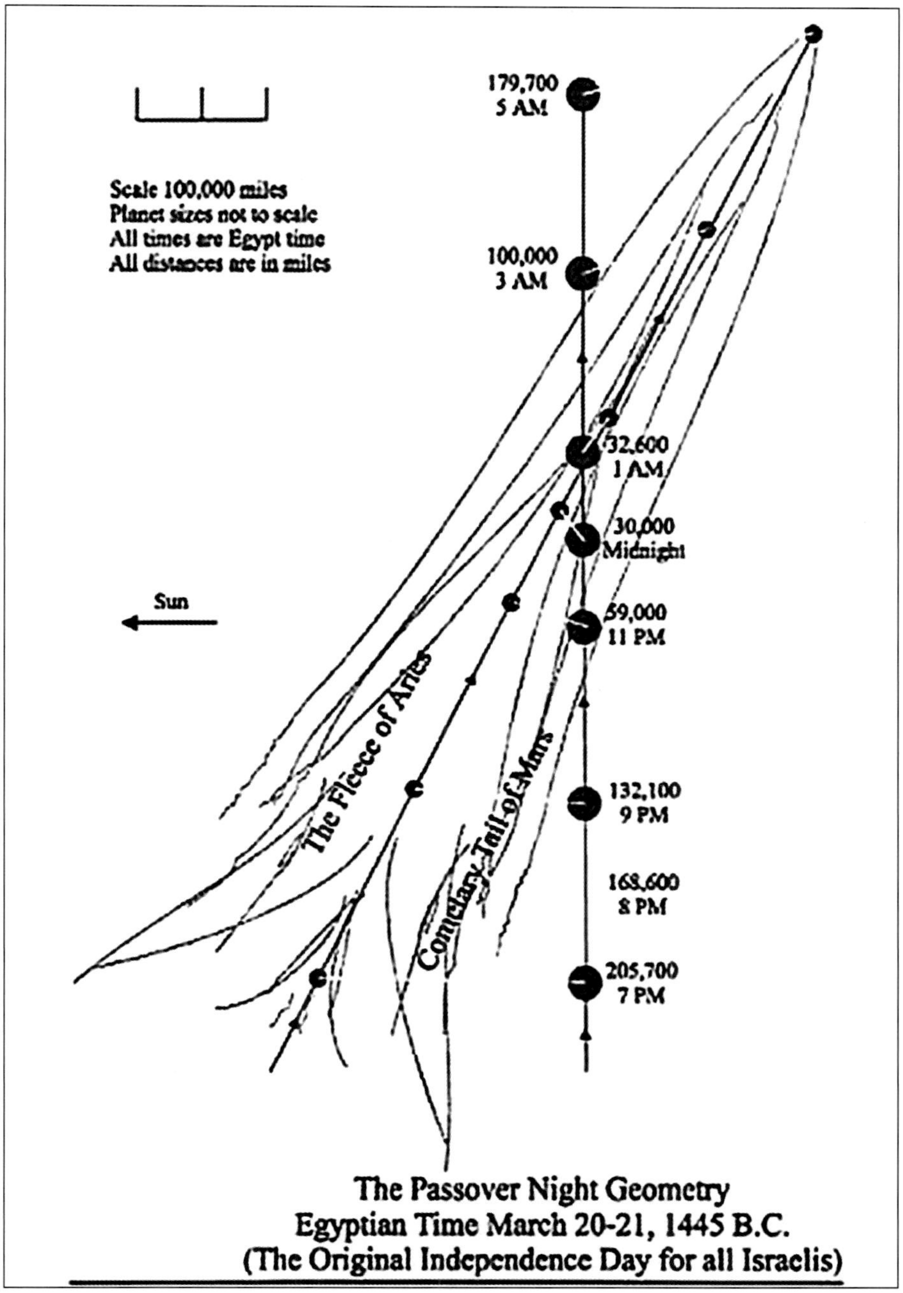

Figure 16 - The Passover Geometry II

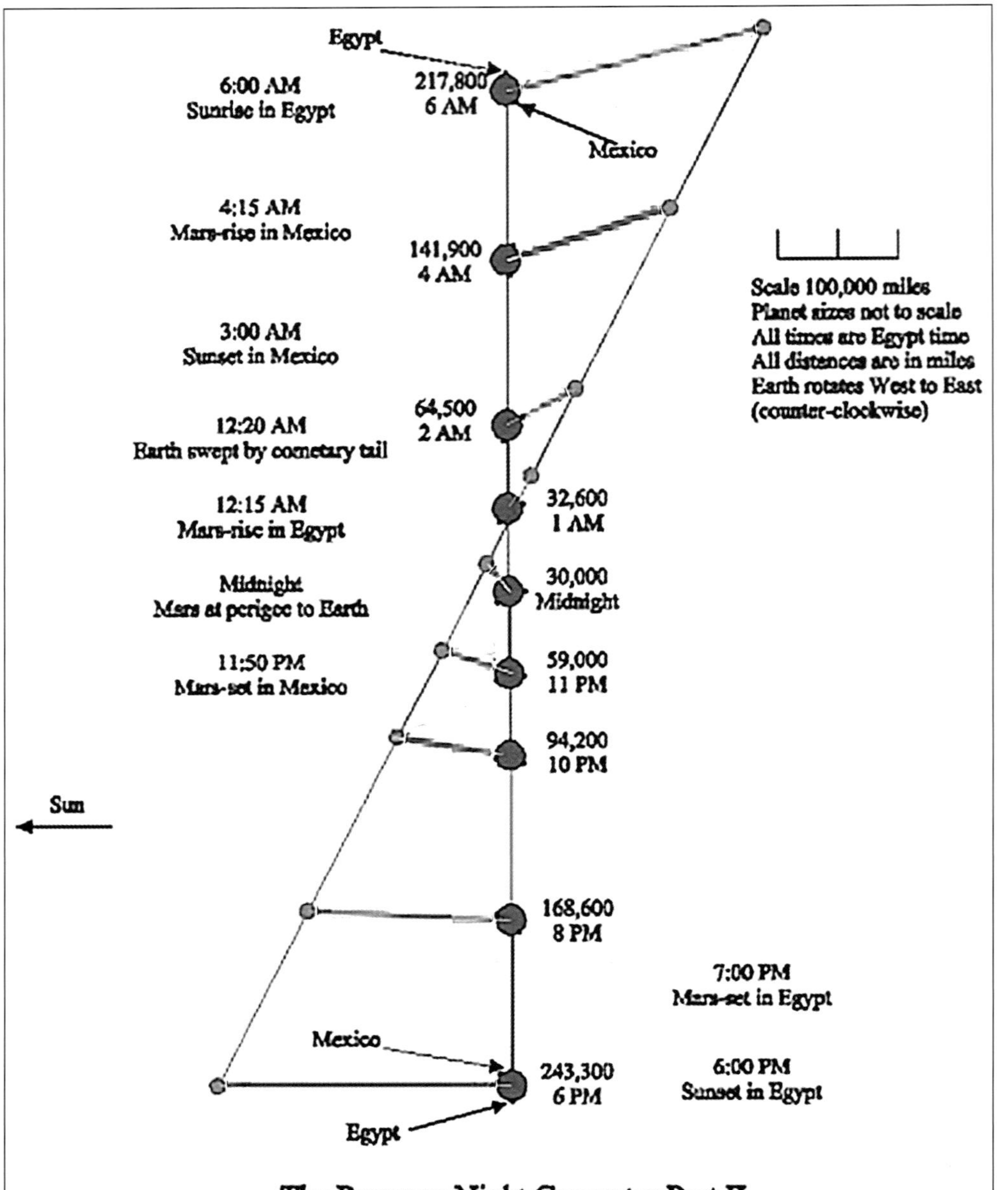

The Passover Night Geometry Part II
Egyptian Time March 20-21, 1445 B.C.
(The Original Independence Day for all Israelis)

Then the Earth shook and trembled; the foundation heaven moved and shook, because he was wroth.
There went up a smoke out of his nostrils, and fire out of his mouth devoured: coals were kindled by it.
He bowed the heavens also, and came down; and darkness was under his feet.
AND HE RODE UPON A CHERUB AND DID FLY: AND HE WAS SEEN UPON THE WINGS OF THE WIND.
AND HE MADE DARKNESS PAVILIONS ROUND ABOUT HIM, DARK WATERS AND THICK CLOUDS OF THE SKIES.
Through the brightness before him were coals of fire kindled.
The Lord thundered from heaven, and the most high uttered his voice.
And he sent out arrows and scattered them; lightning, and discomfited them.
And the channels of the sea appeared, the foundations of the world were discovered, at the rebuking of the Lord, at THE BLAST OF THE BREATH OF HIS NOSTRILS.
II Samuel 22:8-16 (Repeated in Psalm 18:7-15)

As was mentioned earlier, red was the color of the face of Mars especially when seen by Greeks through an Earth atmosphere, clouded with smoke from prairie fires, forest fires and volcanic ash. On the other hand, black was the color of the backside of Mars, as it rotated into its night time posture.

Is it by chance that these two colors, red and black, were the traditional colors of the religion of Baal, the Phoenician belief system. It was also the colors of the belief system of their colonists, in Carthage. They, like many others in their different ways, were first of all, Mars-worshipers. Second in their pantheon was Astarte, Venus. Following those two was the host of heaven, including Jupiter, Saturn, Mercury, the Sun and the Moon.

The Hebrews were otherwise, or at least they were supposed to be otherwise, non-conformists if they followed the messages of Moses and the prophets. The Creator (and His creation), not the planets, were to be venerated. On this particular point, Abraham's faith departed radically from his Chaldean civilization, and from the teachings of his forefathers.

As Mars approached the Earth in a March flyby geometry, its tail followed it. Mars and its fleece looked like an approaching, onrushing whitish, yellowish, glistening pillar. But after it passed the Earth, its tail appeared more like a cloud, from the rear, with its tail visually engulfing its nucleus. This was the pillar of "fire by night" and the "cloud by day" of the Exodus catastrophe.

The Exodus story features both descriptions. It was during just such a destructive flyby of Mars, a "Passover" (of Mars, the angel of death). A careful reading of Exodus chapter 14 is instructive. The appearance of the cometary tail of Mars shifted from a "pillar of fire" to a "cloud".

> And the angel of God, which went before the camp of Israel, removed and went behind them; and the pillar of the cloud went from before their face, and stood behind them. Exodus 14:19

On that night of March 20-21, Mars made a close flyby between the Earth and the Sun; the Moon was at full on the other side, out of the way. Prime time that night was when the Western Hemisphere including the Mayans in Mexico faced Mars and the Sun. For the Hebrews in Egypt, the crisis hour of the flyby was during the night of the Near East.

The Time Table Of The Exodus Passover

The timetable of the Exodus Catastrophe was roughly the following on that famous Passover night. For the Hebrews, it was to be their last night of slavery in Egypt, and that next morning it was Independence Day for all the Hebrews that followed Moses. In space, Mars was advancing on the Earth at a velocity of 30,000 miles per hour.

A MODEL OF ALL OF THE MARCH MARS FLYBYS. All except the Final Flyby were sunny side flybys. Mars, with its cometary tail flying, crossed the Earth's orbit some 30 minutes ahead of the Earth. Its tail swept across Mexico and the Western Hemisphere; within hours it began to sweep the Eastern Hemisphere also.

In Egypt it was night time at the minute of the Mars perigee, estimated at between 30,000 and 40,000 miles. Richter scale 15 earthquakes rattled the bulge spot, the crust of the Caribbean, while Richter scale 13 and 14 earthquakes rattled the flat spot on the opposite side, including the crust of Egypt.

The cosmic scenery was splendid, if one could view it from a neighboring planet. Much structural damage and collapse occurred to Egyptian buildings. The experience was frightful. As Mars rose in the east on March 21, Moses gave the order to get going, now, Now, NOW.

They needed a head start, and somehow Moses knew it. The Hebrews fled the land of Goshen, the eastern part of the Nile delta, as fast as their livestock would allow. It was almost as if Mars was a sign to get going, high overhead in the otherwise dark hours of the night.

The next morning a few regiments of well-armed Egyptian chariots and cavalry set out to recapture the fleeing vassals, who had an eight or ten hour head start. The Hebrews as a group were able to move perhaps 3 or 4 miles per hour; the Egyptian charioteers were much faster.

BUT, about the time of the perigee of Mars, in the middle of the Egyptian night, Thera, an underwater, sub-surface volcano 60 miles north of Crete went into a major league explosion and eruption. It sent ash and cinders all over the Eastern Mediterranean. Next, its walls collapsing back into the Mediterranean Sea, it sent tidal waves 600 feet high sweeping across nearby Anaphi Island. Sixty miles to the south, the north coast of Crete and its Minoan Civilization suffered a 250-foot tidal wave, and the Minoan Civilization collapsed (never to rise again).

Some 450 miles away, on the coast of the flat, low Suez Isthmus, the tidal wave had abated to a mere 75 to 100 foot surge above sea level. The pursuing Egyptian regiments with their chariots and cavalry were taking the low road, and were poorly positioned to cope with this sudden, watery envelopment. Most of the charioteers were engulfed.

Moses, like Noah 1,000 years earlier, somehow foresaw something like this. Hence his indomitable and timely leadership. Like Noah, Moses credited to Ea, or Yawheh for the escape of the Hebrew slaves and their livestock. On the other hand, the Egyptians suffered additionally to the night's earthquakes.

Egypt was downwind from Thera. The next morning ash from Thera fell 10 or 20 inches thick across the delta, and this didn't help matters. The Egyptians blamed Horus and/or their Phoenix bird. The Greeks blamed Ares, and perhaps his offspring, Typhon, Perseus, Gorgon or Medussa.

The Phoenicians attributed this catastrophe to Baal, and sacrificed a special offering, a dispatch of an extra big batch of screaming infants to the fiery furnace of Baal, their idea of a propitiation. It was a rough night in the Near East, and it was even a rougher day in the Caribbean. An approximate time line is as follows.

6:00 p.m. Mars 243,000 miles distant. The Sun sets in March 20, 1445 B.C.E. The incoming Mars did not yet set in the west until 9:00 p.m., and was brighter than the Moon. The Moon was in full phase that night. The Hebrews prepare to celebrate a Seder, and get set to hightail it out of Egypt, with possessions including livestock. Their carts were getting loaded.

9:00 p.m. Mars 132,000 miles distant. Mars sets on the western horizon of Egypt. Earthquakes commence with increasing intensity. The mostly submerged volcano Thera, 60 miles north of Crete, rumbles at an ominously level.

12:00 a.m. PERIGEE. Mars center is at 35,000 to 40,000 miles from the Earth's center. Richter scale 14 and 15 damage the Western Hemisphere, closer to Mars. Scale 12 and 13 earthquakes devastate Egypt, occurring especially severe about midnight, perigee for the angel of the Lord. Electrical damage as well as heavier earthquake sweep across the Western Hemisphere. Thera erupts in a major league eruption. The wind is from the northwest, blowing toward Egypt. Tides 100 feet high sweep many parts of the Caribbean Sea and the nearby Atlantic Ocean.

12:15 a.m. Mars at 30,600 miles, and rises on Egypt's eastern horizon, and is perhaps 25 times brighter than the full Moon. Earthquake and electrical damage wreak havoc in Mexico and the American Southwest. Thera's volcanic crater walls collapse, creating seismic sea waves 600 feet high on nearby Anaphi Island. Moses gives the order to go, now, Now, NOW. Arizona. Putting as many miles as possible between his congregation and Egypt he deems essential. Their route is out of the low lying Suez Isthmus and toward higher ground.

3:00 a.m. Mars is leaving, now over 100,000 miles. Mars-shine is still 20 times as bright as is Moonshine. Earthquake activity is largely over in Egypt. The Hebrew people, a mixed group, moves out at about 3 mph and now is 8 or 9 miles down the road to Elim. Seismic sea waves from There smash 200 feet high across the northern coasts of Crete, destroying the Minoan civilization, which never rebounded.

6:00 a.m. Mars is at 217,000 miles. A smoky, ashy morning as ash from Thera, 10 to 20 inches thick, begins to fall across the Nile delta. The Egyptians arise to assess the damage, and to bewail the dead, and find the Hebrews gone. They are now about 15 miles down the road toward the wells of Elim. Seismic sea waves 150 feet high in places begin to lash and pound the coast of Turkey and the coasts of the Greek mainland. Mars is 10 times brighter than the full Moon.

9:00 a.m. Mars at 301,000 miles. Mars is 7 times as bright as is the full moon. Pharaoh orders some regiments of cavalry and chariots to pursue the fleeing Hebrew vassals, vowing on them bloodshed and a heavier degree of slavery. Thera's seismic sea wave spreads out across the Eastern Mediterranean at 50 to 60 mph, lessening in depth as it broadens. The Hebrews are now weary and some 20 miles on their way to the wells of Elim as Egyptian cavalry and charioteers begin pursuit.

12:00 p.m. Mars at 390,000 miles distant, and showed a disc still somewhat larger and brighter than the Moon. The Egyptian cavalry takes the low road across the Suez Isthmus. Ash and cinders from Thera blot out direct sunlight throughout the Nile Delta, as volcanic ejecta compounds the recovery from the ruin and devastation across the Nile Delta. The Egyptian cavalry and charioteers are perhaps five miles behind the fleeing Hebrews, and the seismic sea wave crashes across the Mediterranean coast of Egypt, now only some 80 to 100 feet high, but still advancing 50 or 60 mph. The Hebrews are 5 to 10 miles ahead of the Egyptian army, and many were wondering about into what kind of a mess Moses was leading them. The seismic sea wave now was about 5 to 10 miles behind the pursuing Egyptian charioteers, still taking the fast, low road across the Isthmus. Show time was about to begin, and the armed forces of Egypt had forgotten to bring their swim suits. Show time was only minutes away.

3:00 p.m. The Hebrews breathe easier, and Moses gives a prayer of thanks to Yahweh. Mars was about 550,000 miles distant, and was a disc slightly smaller than the Moon, but it was still brighter, being more reflective. The Moon would appear on the eastern horizon in three hours, and Mars, departing with its splendid cometary tail in six hours. The momentary safety of Israel was now assured, and the people were breathing much easier. Moses and Joshua now looked like marvelous leaders. It was Independence Day for Israel, a day to be remembered for many reasons. Illustrated in Figures 15 and 16 is the celestial geography of the occasion. Mars and it cometary tail would rise across their eastern horizon in about six hours.

> And the Lord went before them by day in A PILLAR OF A CLOUD, to lead them the way; and by night in a pillar of fire, to give them light; to go by day and night; He took not away the pillar of the cloud by day, nor the pillar of fire by night, from before the people. Exodus 13:21-22

The cometary tail of Mars that night was among the scars of Mars. The erupting volcanoes visible on the surface of Mars were scars of Mars, as were erupting volcanoes on the surface of the Earth, and the consequent fallout of cinders and ash.

Seismic sea waves and huge tidal waves, that day were another of the scars of Mars, as were the earthquakes. The electrical damage that day was specific only to the facing hemisphere of the Earth, which happened to be the Western Hemisphere. The watery destruction that destroyed the Minoan Civilization on Crete also was among the scars of Mars.

It had been the most dramatic of all Passover nights, and Mars passed over. This angel of the Lord was physical, spherical, with a cometary tail of effervescing ices, and had a mass 11% of the Earth's mass. Being the Roman tubulustrium, it was a night of troublesome events. Being the Hebrew Passover, it was Independence Day.

Conclusion

Many scars are physical, and others are mental, or psychological. Story 15 is THE FLEECE OF ARIES; it was one of the scars of Mars, with ices effervescing from its Eastern Hemisphere in the former age and in its former, catastrophic orbit. The "WINGS OF THE ANGEL OF DEATH" were the same visual reality.

The angel of death, a very real angel, "Passed Over" the land of Egypt and lapped our planet. The cometary tail of Mars, story 15, is another new perspective for good astronomers who aspire to become good cosmologists.

There is a logical explanation for the dry river beds of Mars. Rampaging rivers once flowed there, but only for a day. Story 16 is that THOSE RAMPAGING RIVERS WERE CREATED BY A SPRAY OF ICE FROM SPACE, FROM ICY FRAGMENTS OF GLACIS. The spray was across only the Eastern Hemisphere of Mars plus a tiny wedge of its Western Hemisphere.

The icy spray on Mars was thousands of years after the rocky fragment spray by Astra, but not millions of years. That the icy spray was later is proved by some of the Martian craters, some with rampaging rivers flowing in one side of the crater, filling it up, and then flowing out the other side, same volume.

Story 17 is that THE ICE BALL GLACIS FRAGMENTED ON THE EARTH'S ROCHE LIMIT, spraying both planets simultaneously. Thus it is that the ices of the Martian lakes effervesced off into space. Those effervescing ices formed the cometary tail of Mars. Glacis was similar in composition and in size to the ice balls revolving around Saturn and Uranus, except it revolved around Mars.

The most recent ice age of the Earth and the dry rampaging rivers beds of Mars are related, both in cause and in timing. The timing was simultaneous with the closest of the Mars flybys, and with the completion of the construction of Noah's Ark, 2484 B.C.E.

Story 18 is that the shattering of Glacis on the Earth's Roche Limit is also the most logical explanation for THE GENESIS OF THE SHORT TERM ICY COMETS. Such a hefty, recent icy spray is the most logical explanation for the survival of some short term icy comets to this time, 4,500 years later. In addition, it is the best explanation for the distribution of the short term comets in the Inner Solar System. Those fragments maintained much of the energy Mars had in its catastrophic orbit, allowing them to go out to 225,000,000 miles from the Sun.

The genesis of the rampaging rivers of Mars, of the Earth's ice age and of the short term icy comets in space all occurred as products of one event on one day. It was the day Glacis shattered, some 11,000 miles from the Earth's center, with Mars approaching nearby, very nearby.

Thus it is that six planets all have suffered "Little Bangs", nearby fragmentations. Four experienced rocky debris, Neptune, Uranus, Jupiter and Mars, and of these four, Astra's fragmentation was the biggest little bang. Two experienced icy debris, Saturn being one and the flyby scene of Mars and Earth being the other.

Mars, our nearby neighbor, has suffered two fragmentations, one rocky and one icy. Gradualist astronomers have missed the both Little Bangs for no good reason. The Earth has suffered one, an icy one, less than 5,000 years ago, and gradualist astronomers have missed it also.

Story 19 is that THE COMETARY TAIL OF MARS EXISTS IN NUMEROUS ACCOUNTS IN ANCIENT LITERATURES. The cometary tail of Mars was the basis of the Egyptian story of its Phoenix Bird, with fiery feathers flying. It was also basis of the story of the Mayan celestial bird, Quetzacoatl, fiery feathers also flying behind.

It was the basis for the dragon-like writhing tail of the Chinese fire star, Mars. Comparable mental scars have been left with other North American Indian tribes in their lores, as well as with the Vedic Indians of India. All this is in addition to traditions of the Fleece of Aries and the Wings of the Angel of the Lord.

One potentially valuable lesson to be learned is that, based on escape velocities, Mars never has had life as we know it on its surface. There is no water vapor, no oxygen and no growing season. Using the premise of maybe finding ancient fossil life on Mars for a $4,000,000,000 grant for funding a space mission to Mars is a ploy. It is a ruse, unworthy of use by scientists in getting funding for space programs. Ruses often succeed in the short run, but they usually backfire in the long run.

Ruses are not needed; there are plenty of valid reasons for Mars missions, Mercury missions, Venus missions, etc.

The popular Oort model for cometary genesis requires that somehow, icy comets relocated from the edge of the Sun's domain, 2+ light years distant, to the inner regions of the Solar System. It may be popular, but it is without evidence. The more recent Clube-Napier model is somewhat better. It suggests that icy comets were involved and caused Earth catastrophes in the era of recorded history. But icy comets were products of planetary catastrophism, not causes.

In the Exodus catastrophe, it was Mars, not its cometary tail, that did the damage. In the Clube-Napier model, they have the cart and the horse, and in that order. This is much better than the Oort model, but yet has its deficiencies.

Ice balls were delivered to the Solar System by being towed satellites of planets like Jupiter and Saturn, which Little Brother delivered to the Sun. One, Glacis, revolving around Mars, made it into the inner Solar System, co-orbiting with Mars until the day it fragmented.

The closest of all of the Mars flybys set the stage for seven events all at once. They were:

1. The shattering of Glacis,
2. The rampaging rivers of Mars,

3. The sudden rain accompanying Noah's Flood,

4. The Earth's subsequent gathering of ices from space in two ice dumps over its two magnetic polar regions,

5. The genesis of the close in, short term icy comets,

6. The cometary tail Mars once had in ancient times,

7. Monster tides in the Indian Ocean which floated the Ark of Noah into the mountainous terrain of Inner Asia.

That is a lot of achievement for one small 500 or 600-mile ice ball.

Somehow, like Noah and Isaiah, Moses was prescient, and had foresight, omniscience with regard to the future. This enabled him to be the indomitable, foresighted leader that, like Noah and Isaiah, he became.

PERCIVAL LOWELL. Percival Lowell illustrates the power of wishful thinking, a hankering and yearning within the soul for life to be found elsewhere in the cosmos. Many men find the prospect of life elsewhere in the Solar System a fascinating prospect, with or without solid evidence. William Herschel, discoverer of the spin rate of Mars, was among the earliest. Along with Percival Lowell were the Pickerings, William H., Edward C. and James S., and many, many others.

Percival Lowell serves for more than just an example of wishful thinking 90 years ago. As he sat on the edge of the catastrophic World War I, he asked the wrong question, "Is there intelligent life on the planet Mars?" Wrong planet.

Ninety years later, mankind is sitting on the verge of something potentially far worse. Minor league dictators, lacking Christian ethics, are gathering arsenals of major league weapons, missiles, nuclear bombs, chemical warfare packages and biological warfare packages. Some of them look upon Americans, Jews and their own women like donkeys. The first one who triggers the delivery of a missile to Jerusalem or New York City gets to go immediately to paradise, where a harem of 70 attractive, anxious ladies await. It is borderline madness.

The question Lowell ought to have asked is whether or not there is intelligent life on the planet Earth. And if there is, for how long.

With story 19, the readers are 51% of the way to the penthouse of Mars planetary catastrophism.

Modern gradualists have now missed both little bangs, both "recent". The score now is ancient writers 2, modern 20th century gradualists 0.

CHAPTER 8

The Twin Tilts of Mars and The Earth

MARCELLUS;
It faded on the crowing of the cock.
Some say, that ever 'gainst that season comes
Wherein our Saviour's birth is celebrated,
The bird of dawning singeth all night long:
And then, they say, no spirit dare stir abroad;
The nights are wholesome; THEN NO PLANETS STRIKE,
No fairy takes, nor witch hat power to charm;
So hallow'd and so gracious is the time.
HORATIO;
So have I heard, and do in part believe it.

Wm. Shakespeare. [n1]

As recently as Shakespeare's time in England, less than 400 years ago, apparently there still remained the tradition of ancient planetary catastrophes. They even included details about the ancient appearance of a planet - and perhaps its two satellites - "leashed hounds, crouching at his heels." [Perhaps this is a reference to Phobos and Deimos.]

Another question is, "Where would Shakespeare get the idiom of "the port" of Mars?" The American Heritage dictionary says that "port" is an archaic term for the cover of a porthole. How would Shakespeare conceive that Mars was seen through some sort of a covered window, or porthole?

The Hamlet passage talks of the legend that when "no planets strike," it was a wholesome time. The converse is implied, that when planets do strike, it is not "a wholesome time." Horatio, Hamlet's friend, agrees with Marcellus, suggesting that he, too, was aware of ancient catastrophic legends, or traditions.

Introduction

The Earth's rate of rotation is very similar to that of Mars. They are 98.2% alike.

Earth Rate of Rotation 1,436 Minutes
Mars Rate of Rotation 1,477 Minutes
Similarity 98.2%

There are three pairs of twin spins in this Solar System, which was treated in Volume I, Chapter 4, entitled "The Acquisition of Planetary Spin Rates." This acquisition occurred during the primordial period when both the Earth and Mars orbited Little Brother, and were very distant from the Sun. That was the era of the "First Orbit of Mars."

The two spin axis tilts of Mars and the Earth also are very similar. They are 97.7% alike.

Earth Spin Axis Tilt 23.44°
Mars Spin Axis Tilt 23.98°
Similarity 97.7%

These pairs of similarities have been known since the time of the famous astronomer, Frederick Wilhelm Herschel. In 1781, the same year he discovered Uranus, Herschel discovered the axial tilt of Mars. Herschel proceeded to compare the similarity of the Mars tilt to that of our planet. He concluded that Mars had a climate much like our own. In this bit of wistful thinking, Herschel became the forerunner of Percival Lowell.

Traditional astronomers have come up with an inadequate explanation as to why Mars has two tiny asteroid-like satellites. There is a difference between having a bad explanation and having no explanation at all. There has been no explanation at all for the twin spin rates of Mars and the Earth until that one was advanced in "The Recent Organization of the Solar System," Volume I of this series.

Though the twin tilts of Mars and the Earth have been known for over 200 years, to date, traditional astronomers have given no reason at all for this condition, other than "Coincidence." "Coincidence" and "chance" have been somewhat acceptable as reasons; actually they are admissions of ignorance. But ignorance is a fair response. No reason at all is better than a bad reason.

The Earth As A Gyroscope

The properties of a gyroscope first were studied in the early 19th century in Germany by Bohnenberger and in England by Lang. Gyros have properties that seem strange to people unfamiliar with their unique behavior. For instance, if a ball at rest is pushed, it retreats from the force at an angle of 180°.

If it is a spinning ball, a gyro is pushed sideways, 90° to the torque, at right angles. A torque applied to a gyro also will cause tiny change in its speed of rotation. That change in spin rate is always minute, and however little, it is always an increase. Thus, it requires a long series of many repeating torque's to create a significant increase in spin of a gyro.

Foucault, a French physicist and a showman, demonstrated the famous Foucault Pendulum in Paris in the 1851. It created a circus-like atmosphere, to the amazement and delight of the citizens of Paris. Then a curiosity, modern society has since found numerous practical applications for gyroscopes.

Before understanding the cause of the twin tilts of these two spin axes, it is necessary to identify Mars and the Earth for what they actually are. They are a just a pair of gyroscopes, orbiting at tens of thousands of miles per hour in space. The smaller gyro has an equatorial diameter of 4,212 miles and a spin rate of 1477 minutes. The larger gyro has an equatorial diameter of 7,927 miles, and rotates a bit faster, in 1436 minutes.

A gyroscope is an object that spins about an internal axis. In industry, usually they are made of a metal such as steel. But gyros also may be made of glass, wood or plastic. A plastic top, spinning on a table, is a gyro, as is a merry-go-round at a carnival, a rotating wheel on a car or motorcycle or a bicycle, the armature of a motor, etc. Even a spinning pair of dice are gyros. These gyros all function within the Earth's field of gravity, which is always a significant force.

In the Solar System, in space, gyros function independent of the Earth's gravity, and do not necessarily have crusty surfaces. Jupiter is a rapidly-spinning gyro composed of hot gases, a 87,000-mile diameter and a 590-minute spin rate. Saturn is a gas gyro also, with a diameter of 71,500 miles and a spin rate 602 minutes. Both of these gas gyros have no crusty, solid parts at all. (Like Mars and the Earth, Saturn and Jupiter also have twin spin rates).

Torsion

Torsion is the process of applying two opposing forces to an object where the forces are not "in line." This 8th chapter involves a study of "torsion past" on the gyro Earth and on the gyro Mars. When there was a flyby, the angle between the two planets was a moving angle, and the angle to one was the same angle to the other. In a flyby scene, the gravity of Mars put a torque on the Earth, and at a similar angle, the gravity of the Earth put a torque on Mars.

THE BOMARC MISSILE. On the Earth, so far, all but one gyro has been solid. There is one exception to solid gyros of which we are aware. It was the Bomarc Missile, a fluid-fueled projectile. It was a projectile in flight, like a spinning bullet or a rotating artillery shell. With respect to its explosive charge, its casing and its guidance system, the Bomarc missile functioned as a solid gyro.

But when spinning, it kept missing the target, badly. It turned out that its liquid fuel did not function like normal, solid components of a gyro. The molecules in fluid gyros function according to an entirely different set of principles than do the molecules in solid gyros.

Liquid Gyros: A New Field For Research

When a torque is put on a gyro, its response is to "precess." If it is a fluid, one effect is that its spin axis relocates on the gyro, as well as toward surrounding space. This is not true of solid gyros.

An effect is that its spin axis shifts, not 180° away from a push, or torque, but rather, it shifts 90°, which is perpendicular to the push, or torque. In the case of gyros rotating in the counter-clockwise mode, like Mars and the Earth, shifts of tilt were 90° to the right. For clockwise spin, the shifts are 90° to the left.

In solid gyros, when they precess, the entire gyro shifts as a unit. All molecules retain their pre-torque partner molecules, and all molecules shift in lock step.

It has been discovered that when fluid gyros precess, the molecules therein do not shift in lock step. Rather, all molecules retain their original velocities, but they change places with other molecules. Instead, fluid molecules relocate.

For instance, the motions of fluid molecules, when flowing in pipes around 90-degree ells, are instructive. When flowing around a 90-degree turn, do some of the molecules increase in velocity and others decrease so as to make a smooth turn? Do the molecules play "crack the whip" and some in making a turn go faster than others? No, not at all.

In the case of water flowing around a turn in a pipe, all water molecules want to maintain their original, uniform velocities. To do this, going around a 90-degree turn, EACH MOLECULE MUST SHIFT POSITION. They spiral around the 90-degree turn, and in so doing, each molecule ends up surrounded by new molecules. In this way, the laws of energy and momentum conservation are met.

But for solid gyros, all molecules maintain lock step with each other. This is why the Bomarc missiles required a different set of equations for its liquid fuel.

Figure 17 illustrates the motion of molecules, flowing in a pipe, making a 90° turn.

Fluid gyros, and the motion of their molecules, essentially have yet to be studied, and especially in an environment independent of the Earth's gravity.

Figure 18 illustrates what is expected, from engineering principles, in flow patterns of magma inside a planet's crust. For molecules of magma, whatever their individual velocities before the spin axis shift, those same velocities will be maintained both during and after the flyby event.

Molecules near the spin axis have low velocities. Molecules in the Earth's magma just under the equator have high velocities. For instance, molecules under the equatorial crust of the Earth rotate at velocities of 1,030 mph. Other molecules within a mile of the spin axis rotate at velocities of .3 mph or less. Liquid molecules inside Mars or the Earth display a wide variety of velocities. Their velocity depends on the spin rate of the gyro and their distance from the spin axis.

In the case of the Earth or Mars, complexities ensue from the fact that bulges try to form in the liquid mantle, stressing the under side of the crust. The bulges form quickly because the meat of the approach of Mars is only four hours or less long. And the meat of its retreat is of similar duration. (Mars advanced about 120,000 miles upon the Earth in a four hour span of time.)

Figure 17 - Molecules of a Fluid Twisting Through a Turn

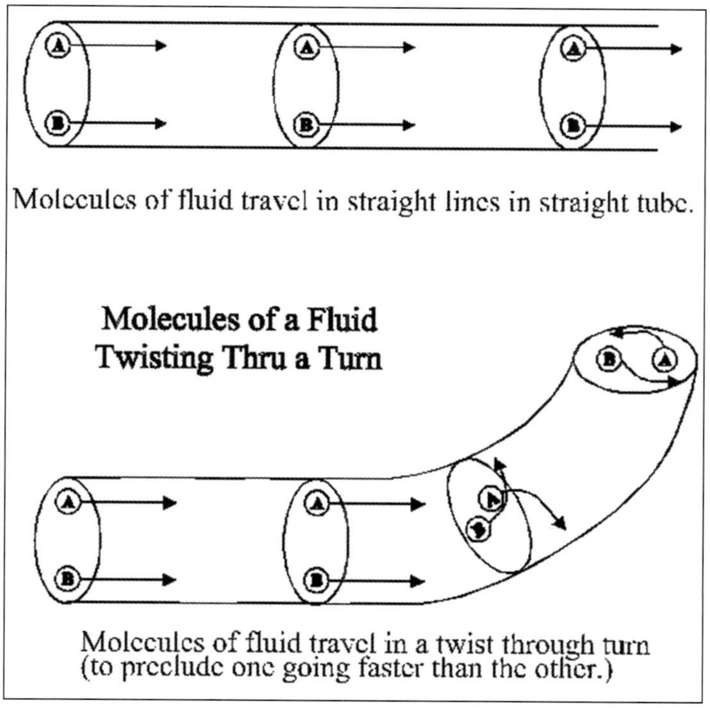

Figure 18 - Molecules of Fluid Magma in Flyby Conditions

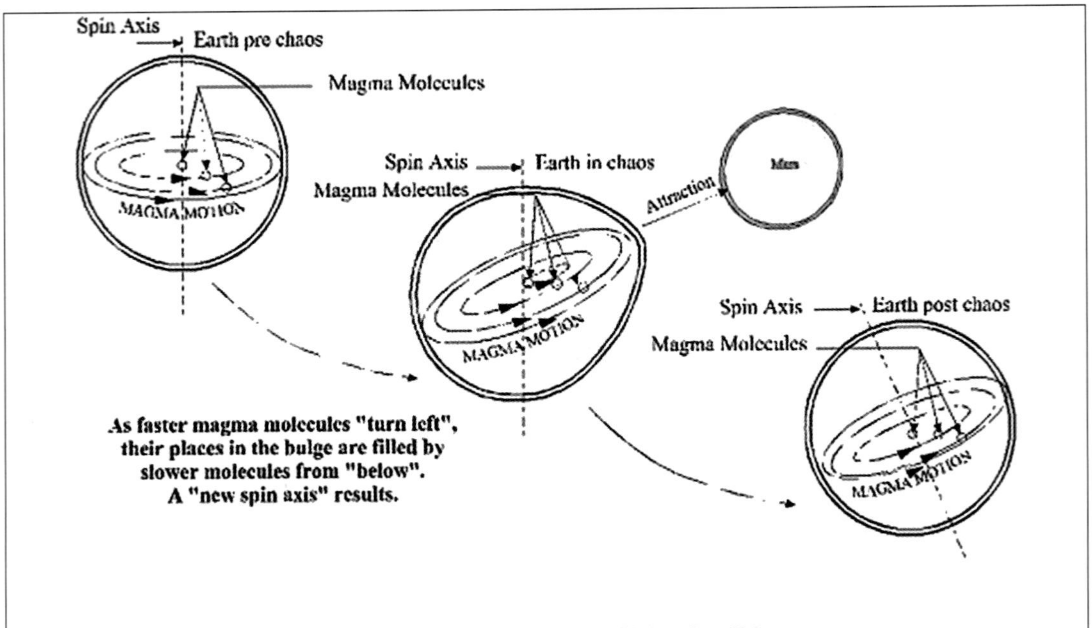

Molecules of Fluid Magma in Flyby Conditions

A bulge forms within the Earth, facing Mars; the bulge stresses the crust just above the bulge. The "flat" formed directly opposite the bulge also stresses the crust. Both the bulge and the flat moved at rapid rates due to two reasons, (1) to the rapid approach of Mars, and (2) to the Earth's internal spin rate.

That rate of bulge increase is eight fold for each halving of the distance between Mars and the Earth. It is the tidal ratio, mass over distance CUBED. Thus the pressure under the bulge on the Earth at 30,000 miles was 64 times (8 x 8) as great as that subcrustal pressure at 120,000 miles, only three hours earlier.

The bulge rapidly expanded facing Mars during the close approaches of Mars, but the bulge also relocated 30° of longitude per hour due to the Earth's spin. Both bulge-moving functions occurred simultaneously. Next, the bulge contracted just as rapidly as it expanded when Mars departed and the flyby was over.

The Earth is part fluid gyro, magma, and part solid gyro, the crust. The crust is 0.006% of the total Earth mass. The fluid magma and mantle are about 99.39%. In addition the Earth mass is about 0.00002% water, its sloshing, uncapped, tide-traveling oceans. This is merely an introduction to the complexities involved on a planet like the Earth with a resisting, but flexible crust. That crust was "tortured" - pressured with suddenness, rinsed from on top by oceans and flexed from underneath by fluid magma.

The elastic crust of the Earth has an elastic-plastic threshold. When that threshold is exceeded by sudden pressures from subcrustal magma tides, the crust shears, or rifts, or tears, or up thrusts. Hours later, it collapses back, only in part after the uplifts, shears, rifts and tears.

In short, the result was sudden crustal deformation. The formation of mountain ranges tended to be in arcs, which followed a great circle. Uplifts were followed by a partial collapse of both mountain uplifts and intermontane basins, under which magma had evacuated. The behavior of the Earth's thin crust was a response to the need for sudden, temporary relief from equally sudden internal distresses deep under our planet's crust.

A solid gyro precesses by moving its plane of spin. That is, in a solid gyro, the "North Pole" always remains in the same place on the spinning surface. It is not so on a liquid gyro. A liquid gyro precesses by the molecules moving so that the "North Pole" shifts to a different "place" within the liquid, and hence a different place on the thin external crust also.

The Earth is two gyros, one is solid one, the crust surrounding the liquid gyro. The liquid aspect dominates the solid one, 99:1. This means that not only does the tilt of the spin axis change relative to the ecliptic. The equator and the two flat spots on the crust with zero rotational velocity (the North and South Poles) also relocate. [n2]

Yet another aspect is that the crust is mostly (72%) covered by oceans. Salt water tides were not capped, and were not as easily constrained as were subcrustal tides. The oceanic tides overflowed, creating vast, sudden layers of strata, usually on the continents where silts, sediments and rocks could be first entrained.

When the waters drained off, the sediments became strata, sometimes rich in marine fossils such as fossil silted in fishes. And the strata often was rich in terrestrial fossils such as dinosaurs, trees in petrification, etc.

For individual molecules in the subcrustal region, under flyby conditions they began to surge around inside the crust. The sudden, new bulges, and other molecules flowed across the two suddenly enhanced polar flat spots. Meanwhile the bulges were changing location, on an east to west, due to the west-to-east rotation of the Earth.

For individual molecules, as they traveled around the bulges and away from the flat spots, molecule positions shifted in great catastrophic surges, while yet maintaining original velocity. The crust was tortured both from within (surging magma tides) and from without (vast, sudden oceanic tidal swells.)

The crust under flyby conditions, especially under the closer flybys, came under great stresses, especially where it was thinnest and weakest. As was just mentioned, the crust is elastic, but there is a threshold, or limit to its flex. When that threshold was exceeded, mountain ranges were thrust up as the crust split. Crustal deformation is just one of the responses of the crust to flyby torture. It is merely a response to sudden internal distress. Immense external oceanic tides occurred simultaneously.

Great bleedings occurred on the Earth's surface; geologists call them basaltic traps (but without understanding their origin.) One such basaltic trap, the Deccan Trap, outflowed suddenly over 160,000 sq. miles of Central India. At the same time, oceanic surges came into play. In places, oceanic fish were fossilized in Central India, along with various terrestrial vegetation, as cool salt water, entrained silts and sediments, and hot magma all mixed.

For the Pacific Northwest, the Columbia Plateau is another basaltic trap, perhaps the third largest on our planet. It is intermontane, and extends from the edges of British Columbia to the edges of Utah and Nevada. Eastern Washington, Eastern Oregon and much of Southern Idaho were engulfed in this original basaltic outflow. Lava outflowed up to 70 mph at the time.

Arizona is marked by many smaller "piles of basalt stone", erupting through smaller rips in the crust. Like crustal deformation and oceanic tides, basaltic traps were another mechanism to assist the Earth in badly needed relief of its sudden, internal distress.

Evidence indicates lava flows forming the Deccan and the Columbia Plateau flowed at velocities up to 70 mph as they bled out onto our planet's surface. This is merely a reflection of suddenness and the intensity of the gravitational crisis during the crisis flyby hours.

In addition to the planetary gravitational crisis, there was a second crisis, one for the Earth's geomagnetic field. The two planetary magnetic fields of Mars and the Earth briefly UNITED in space, Earth's north pole field and the Martian south pole field.

Hours later they were quickly torn apart, leaving both planets with "paleomagnetic" polarity reversals. The geomagnetic north pole and the geomagnetic south pole of the Earth switched places within a matter of hours. When a basaltic trap flowed, or a volcanic eruption was renewed, the magnetic polarity of that hour was recorded for posterity (us) in the cooling iron in the virgin lava flow. When the next eruption occurred, that polarity reversal also was recorded for posterity (20th century geologists.)

This chapter merely touches on the edges of the complexities involving the Earth's responses to Mars flybys. More important to this chapter is what happened to the Earth's and the Martian two spin axes, and how they attained their two twin tilts. One is at 23.44° and the other is 23.98°.

The sum of the relocation of the molecules of magma produced the spin axis precession. As each molecule was forced to find a new path, the average plane of spin changed. The new Earth's spin axis was simply the center line about which the new average of the molecules of magma spun. So, in an over all sense, it was rather simple in consequence, even if it was complex in its process.

The Earth and Mars suffered spin axis shifts in free (frictionless) space. This is not like a top rotating on a table, suffering a sudden torque or two. In our mundane world, in the environment of the Earth's gravity, the table top applies friction, slowing the rotating top. In the Mars-Earth Wars, there was no table top to slow the process down, or to exert a drag.

Mixing Silanes, Carbides And Water

During Mars flybys, much blending occurred in the magma's chemistry due to sudden, volatile mixing of magma molecules. Sometimes the blend in the mix was (a) silanes and (b) carbides in the presence of (c) water. Carbides are compounds of carbon and another metal. Silanes in the magma are silicon hydrides, complex compounds of silicon and hydrogen, of the general formula Sin H2 n+2.

When they mix, there is a chemical reaction, producing new compounds. The silane's silicon abandons the hydrogen in favor of binding with oxygen. That reaction produces (1) silica, (2) silicon dioxide, (3) more crust. Water's oxygen ions also abandon hydrogen to bind with silicon, more silica yet.

The abandoned hydrogen ions wandered around, like new divorcees, along with some newly wandering carbon ions, that also were new divorcees. As leftovers from previous marriages, they also considered uniting in the presence of (a) high pressures and (b) high temperatures. They decided to go ahead and to unite in a second go around at pressures similar to a cracking plant. It was simple. Those left over ions remarried (polymerized) and became viable, virgin hydrocarbons, in pools, also known as petroleum pools.

PETROLEUM POOLS. The primary result of this kind of mix is more silicon dioxide, which is silica, which is more crust, and a thicker crust. The byproduct is the polymerization of hydrogen ions and carbon ions into hydrocarbons, which is done at temperatures similar to cracking plants. The byproduct is hydrocarbons ... petroleum ... which gathered in pools. [n3]

Those chemical mixes resulted in "interesting" chemical reactions. To modern industry, the byproduct is of more interest than is the primary product, plain old silica. The byproducts gathered to form pools of petroleum at the upper side of the mantle. At the present, neither oil company executives nor petroleum geologists have yet to understand the genesis of pools of petroleum (Hunt and his listeners excepted).

For petroleum geologists, an understanding of recent planetary catastrophism might result in drilling fewer dry holes, in increasing discovery of petroleum pools, and in producing surpluses, good for increasing corporate dividends. Planetary catastrophism, Mars flybys, is (are) what created the high velocity mixing of the silanes, the carbides and water at the molecular level.

The Earth has a polar radius of 3,950 miles, of which an average of 10 to 12 miles is crust. Approximately 99.4% of the Earth's mass is hot magma and mantle, below the thin crust. Such was the potential of chemical reactions in the mantle during Mars flybys in fluid gyros. What happened to the Earth happened to Mars also. Petroleum may well be pooled under the thick crust of Mars.

VENUS. Venus has an even thinner crust than the Earth. By this same carbide-hydride-mix theory, and including Mars flybys, it is probable that Venus also experienced much internal mixing at the molecular level. Hence it is probable that within and just below the crust of Venus also lie large pools of hydrocarbons, or petroleum.

Some of that petroleum apparently has found its way through its thin crust and to the surface of Venus, and then into its atmosphere. Atmospheric pressure, from a thicker atmosphere, at the surface of Venus is 90 times what it is at the surface of the Earth.

On Venus, 97% of its thick atmosphere is carbon dioxide. The Earth's thinner atmosphere is .00035 CO_2. These facts are supportive of the carbide-hydride mix theory for creating silica and hydrocarbons for Venus as well as for the Earth. Hence it must be suspected that the thick carbon dioxide composition in the atmosphere of Venus is just one more of the many scars of Mars. Petroleum pools on the Earth also are scars of Mars flybys.

Spin Axis Shift And The Final Flyby

Four aspects should be kept in mind to understand why the Mars gyro and the Earth gyro ended up with similar spin axis tilts. They are:

1. RECIPROCAL TORSION. Mirror angles occurred between the two planets regardless of their velocities.

2. REPEATED TORSION. In Volume III, a foundation will be laid that Mars flybys occurred once every 54 years. March Passovers were always followed by October case flybys. In 5400 years, there were 100 flybys, and 100 instances of repeated torsion to reset the spin axis tilt.

3. ALTERNATING TORSION. Evidence points to the first original Mars flyby as being a March case. Thus all odd-numbered flybys were March "Passovers" and all even-numbered flybys were October cases. This model of planetary catastrophism specifies precisely this. Alternating torque's are essential to the

maintenance of planet orbit resonance's.

Mars is modeled as having been in 6:1 orbital resonance with Jupiter, in 1:2 resonance with the Earth, and in near 15:1 resonance with Saturn. In addition, but of less consequence, Mars was in 1:24 orbital resonance with the Moon. Yet another orbital resonance was Venus and the Earth in a virtual 8:5 orbit resonance 1:.626).

4. SOFT, FLUID GYROSCOPES. Fluid gyroscopes produce volatile spin axis shifts, much more so than do solid gyros. The Earth, with 98.2% of its mass below the crust, is a fluid gyro. The responses of fluid gyros to torque's in space or on the Earth have yet to be studied by scientists.F4

There are various indications in ancient literature of ancient spin axis shifts, and the resulting changes. This was the reason ancients constructed sun dials, sun spirals, sun caves, etc. During some flybys, the shadow of the Sun at zenith (high noon) lengthened measurably, and the cardinal directions shifted. During other flybys, the shadow of the Sun at zenith shortened measurably. And once again, the cardinal directions shifted a few degrees.

The champion of the Hebrew cosmologists was the prophet Isaiah. He described the Final Flyby in some detail. Some aspects were crustal disturbances. Some were electrical disturbances, including interplanetary discharges. One involved a recorded spin axis shift.

Central to his description was a shift of the shadow of the Jerusalem obelisk. No doubt records were kept of the shadow's length on the floor at zenith, at high noon. Note in the following citation "by which degrees it was gone down." Isaiah 38:8b.

Isaiah had experienced an earlier shift as a child (in 756 B.C.E.). In their national memory, it had been different; it involved a lengthening of the shadow. This shift in 701 B.C.E. was different; it was a shortening of the shadow. This is why Isaiah is quoted as having said "the shadow of the SUN "RETURNED" TEN DEGREES". Below, the spin axis shift aspect of Isaiah's account is cited.

The Shortened Shadow Of The Sun Dial Of Ahaz

And this shall be a sign unto thee from the Lord, that the Lord will do this thing that he hat spoken; behold, I will bring again the shadow of the degrees, which is gone down in the Sun dial of Ahaz, ten degrees backward. So the Sun RETURNED ten degrees, by which degrees it was gone down. Isaiah 38:7-8

The Sun dial of Ahaz is somewhat deceptively entitled. It probably was an Egyptian obelisk, a spire like the Washington Monument. Probably, it was a part of the package of benefits for King Solomon when he accepted the Egyptian princess (with her retinue) as one of his wives, in the 960's B.C.E.

Egyptian engineers constructed and manufactured many obelisks in Egypt. The higher end of the range was 100 to 115 feet, as is measured from still existing ones. The shorter range was likely from 50 to 80 feet high. We have no way of knowing whether the Jerusalem Obelisk (the sun dial) was in the high range, mid range or shorter range of their Egyptian productions. It is assumed the obelisk had a height of 75 feet, a height in the mid range.

It was entitled "the sun dial of Ahaz" because around its base were calibrations according to the current path of the Sun's shadow. Recalibrations were needed on the ground, where the shadow was measured, after every spin axis shift. One had occurred In October, 756 B.C.E. Another more severe one occurred in 864 B.C.E. Apparently there was a recalibration of the base after each flyby. The last one was done during the reign of Ahaz, thus "Ahaz' sun dial." It was essentially a recalibration by Ahaz' mathematicians.

In the account of Isaiah, the shift was reported, and translated into English as "10°" and in the direction of a shortening of the shadow, at zenith, at high noon. The phrase "10°" occurs only one other place in the Bible, in the Psalms.

The Jews lost their Hebrew language while in the Babylonian Captivity. Probably, they exchanged their ancient standard of measures for the more widespread, more modern Chaldean standard of measures. So, after they returned, two or three generations later, the precise meaning of the ancient Hebrew standards of measure had been lost.

So, nobody knows how far "ten degrees" was. It was ten of something. The Hebrew word translated "ten degrees" as "ma'alah". Its definition from Strong's Concordance is:

> 4609 ma'alah, elevation, i.e. the act of journey to a higher place, ... step, a climactic progression ... things that come up (high) degree, deal, go up, stair, step, story.
>
> 4608 ma'aleh, elevation, activity or platform, ascent, before, chiefest, cliff, that goeth up, going up, hill, mounting up, stairs. [n4]

"Ma'alah" is derived from "ma'alah" (4608) and that from "alah" (5927.) The context of these words is measuring in the vertical direction. Surveyors (then and now) measure in the vertical direction by first measuring accurately in the horizontal direction. Then the angle is measured precisely (trigonometry). Thus, in a sense, "degrees" as a translation into English is somewhat on target; degrees do relate to angles.

The studied conclusion of this staff is first, that the height of the obelisk, or sun dial of

Ahaz, (or Solomon), whatever its height may have been, was between 8.5 and 8.6 times greater than the distance the shadow shortened.

Second, the Hebrews used five kinds of cubits. If the "short cubit" was used here as their standard of measure, a "degree" was 10.7 inches, or three palms. Ten short cubits were 107 inches, or 8.92 feet. See Figure 19.

Figure 19 - The Shift of the Shadow of the Sun Dial of Ahaz March 21, 701 B.C.E.

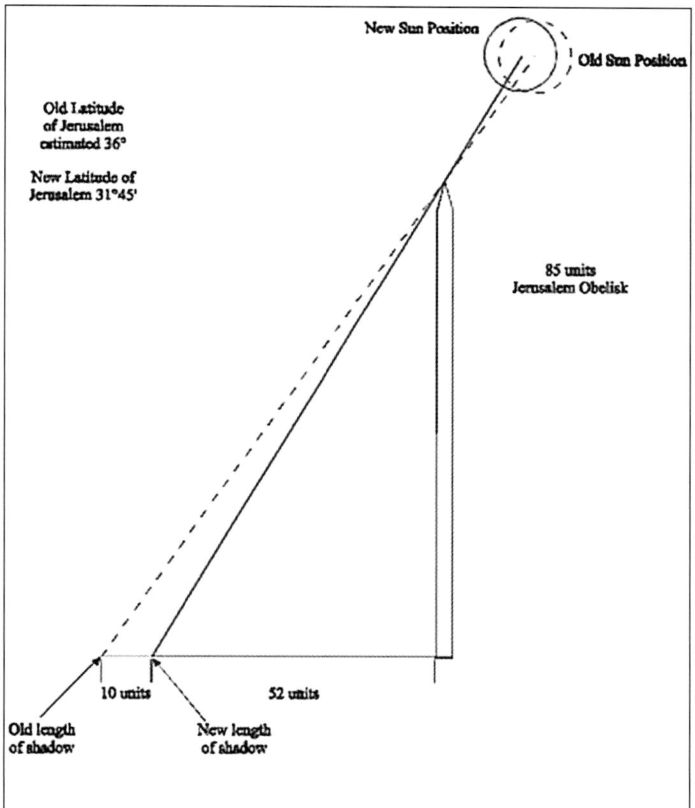

The Shift of the Shadow of the Sun Dial of Ahaz March 21, 701B.C.E.

The obelisk was approximately 8.55 times as high as the shadow shortened. 8.92 feet x 8.55 = 76.3 feet. It is suspected that the obelisk of "Solomon's Egyptian queen" was 76 feet high, +/- 2 feet.

Next it is then concluded that the distance of shortening of the shadow was 8.92 feet, or 107 inches, also 10 short cubits, which also was 20 tophechs or 30 palms. (70,000,00 palms approximated one polar radius). See Figure 19. But first, other shifts in direction of many ancient foundations of Mediterranean temples merits a review.

The Orientation Of 30 Foundations Of Ancient Temples Of The Mediterranean Region

The data from Isaiah about a shift in the shadow of the Jerusalem Sun Dial is of literary quality. It also has some scientific quality. Isaiah was an eye witness. But there is another source of data, which is of modern scientific quality.

Dr. Rafael Kazmann, engineering professor, now retired, at LSU has made an analysis of other studies of orientations of foundations of 30 temples in the Mediterranean region. One of his sources, Dr. Asher Kaufman, discovered the foundations of the First Temple and the Second Temple in Jerusalem.

The first foundation of the First Temple was designed about 965 B.C.E. A study was made of the foundation of the First Temple, Solomon's Temple, in Jerusalem and it is found to be oriented 6° north of the true east according to the cardinal directions of the modern our age.

Kaufman also found, and measured the orientation of the Second Temple, Herod's Temple, designed in the early part of the first century B.C.E. The foundation of the Second Temple matched precisely modern cardinal directions.

> After many years of field work, Kaufman, a member of the physics faculty of the Hebrew University in Jerusalem whose hobby was archaeology, finally established the location of the first (Solomon's) and the second (Herod's) temples and calculated the orientation of each. Both temples were built on the same spot, one set of foundations overlying the other with the same location for the holy of holies. In response to a letter of inquiry, he stated that the first temple was oriented 6° North of East and the second temple was oriented exactly East-West. He estimated his error of measurement to be about 0.2°, more or less. [n5]

Kazmann also cited the conclusion of G. E. Wright, who excavated Shechem.

> It is clear now that the Israelite structure's exterior walls had simply followed the outline of what remained of the Late Bronze Age temple 2. Yet the first and most inexplicable fact about temple 2 is that the axis shifted. Whereas temple 1 was oriented 28° South of East, temple 2 was oriented 33° South of east.... Why the

axis was shifted 5° farther south is an unsolved mystery.... A rather striking parallel to the structures unearthed at Shechem is found in level VII of Alalakh in Northern Syria dating from the Seventeenth or early Sixteenth Century. [n6]

What has happened to the cardinal directions in the Near East during the first millennium B.C.E.? Before giving our conclusion, it is appropriate to cite the 30 temple foundations which Kazmann listed in the Mediterranean region, two of which were in Asian Minor.

Table X - The Orientation Of Ancient Mediterranean Region Temples [n6]

PLACE	DEITY	ORIENTATION 1	EST. DATE
GREECE			
Tegea	Minerva	1 D 23'30" S of E	1650 B.C.E.
Rhamnus	Themis	1 D 29'46" S of E	1092
Rhamnus	Nemesis	1 D 24'50" S of E	787
Athens	Minerva	9 D 05' S of E	2020
Athens	Nike Aptersos	5 D 43'17" S of E	1130
Athens	Esculapius	5 D 33'00" N of E	?
Olympia	Juno	3 D 16'02" N of E	1445
Olympia	Jupiter	7 D 22'14" N of E	790
Corfu	Kadarki	4 D 39'35" S of E	875
Lycosura	Despoina	2 D 11'14" S o E	650
Megalopolis	Jupiter Sot.	9 D 42'28" S of E	605
Thebes	Cabeiron	6 D 27' 45" S of E	1160
Mycenae	Unknown	6 D 40' 20" N of E	?
ITALY			
Croton	Juno Lacina	2 D 34' N of E	?
Metapontum	Unknown	6 D 57' S of E	?
Paestum	Neptune	3 D 09' S of E	?
SICILY			
Girgentl	Juno Lacinia	6 D N of E	690
	Hercules	0 D N of E	470
	Castor	4 D 00' N of E	400
Selinus	Unknown	4 D 52' S of E	795?
	Unknown	5 D 35' S of E	?
	Unknown	5 D 40' S of E	?
	Unknown	6 D 18' S of E	?
	Unknown	6 D 40' S of E	?
	Unknown	7 D 21' S of E	?
Syracuse	Minerva (?)	0 D 42' N of E	?
	Diana (?)	1 D 45' S of E	?
	Olyumpeicium	7 D 26' S of E	?
ASIA MINOR			
Ephesus	Diana	5 D 21' 05" S of E	715
Jerusalem	Jehovah (God)	6 D N of E	925

From the data concerning the shift in orientation during the Flyby of 701 B.C.E., it was concluded that Jerusalem shifted south about 4°, or about 275 miles. Or, a new equator shifted northward toward Jerusalem in that same amount. And the North Pole shifted some 275 miles toward Alaska, the Yukon, British Columbia and Seattle, putting them into somewhat more northerly climates. Europe and Asia Minor shifted southward also 4°.

When temples were built, it was customary that the foundation be aligned with the cardinal directions as of the time of design. The foundation of the First Temple (Solomon's) was designed about 965 B.C.E. and was built during the ensuing fifteen years. Raphael Kazmann quotes Kaufman who discovered the foundation of the First Temple. It is offset 6° from true east of the modern era.

From data to be presented in Volume III, it is known that there were five flybys in the era 965 B.C.E. to the Final Flyby of 701 B.C.E. There were two March case flybys in 809 B.C.E. and in 917 B.C.E. Both seem to have been light, and their effect would have been to pull Jerusalem northerly.

There were October flybys in 864 B.C.E., coinciding with the Trojan War, and in 756 B.C.E., coinciding with the prophetic careers of Jonah, Joel and Amos. Josephus recorded some of the damage done to the First Temple structurally during the flyby of 756 B.C.E.

> ... in the mean time, a great earthquake shook the ground, and a rent was made in the temple, and the bright rays of the sun shone through it, and fell upon the king's face, ... [n7]
>
> And before the city, at a place called Eroge, half the mountain broke off from the rest on the west, and rolled itself four furlongs, and stood still at the east mountain, till the roads, as well as the king's gardens, were spoiled by the obstruction. [n8]

The catastrophe of October, 756 B.C.E. seems to have been middling in intensity. The flyby of October 864 B.C.E., coinciding with the Trojan War, was a megacatastrophe, as was the Final Flyby of March 701 B.C.E. A megacatastrophe is one in which the two planets were within 30,000 miles of each other at perigee.

This analysis is that the flybys of 864 B.C.E., 756 B.C.E., and 701 B.C.E. shifted the North Pole away from Asia Minor and Europe, and toward Alaska and the Yukon. Israel shifted closer to the equator, and its latitude lowered. On the other hand, the flybys of March 809 B.C.E. and 917 B.C.E. shifted the North Pole toward Asia Minor and Europe, and away from Alaska and Canada.

The data from Kazmann and Kaufman indicate Jerusalem shifted southward 6°, or 410 miles as a consequence of the combination of all of the last five flybys. If the 701 B.C.E. event shifted Jerusalem south by 4°, then the combination of the other four shifted

Jerusalem south by another 2°, totaling 6°.

On the basis of Kazmann's data, the conclusion here is that the latitude of Jerusalem slipped south by a total of 6° between 965 B.C.E. and 701 B.C.E. During this time span there were five Mars flybys, three of the March case and two of the October case.

If this interpretation of Kazmann's data is correct, the latitude of Jerusalem shifted southward by some 400 miles. Simultaneously the North Pole (spin axis) shifted some 400 miles in the Arctic Basin. If so, the North Pole net shift was away from Scandinavia and toward Alaska, the Yukon, British Columbia (and Seattle.) As Jerusalem shifted southward 6°, or 410 miles, Seattle, Vancouver, Anchorage, Fairbanks and Point Barrow shifted northward a similar distance and their climates became cooler.

BETWEEN 1450 B.C.E. AND 965 B.C.E. There was a spread of other catastrophes between the time of Moses (1450 B.C.E.) and the building of Solomon's Temple. They occurred in October of 972 B.C.E., 1080 B.C.E., 1188 B.C.E., 1296 B.C.E. and the famous Long Day of Joshua, 1404 B.C.E. Others occurred in March of 1025 B.C.E., 1133, B.C.E., 1241 B.C.E., 1349 B.C.E. and of course the Exodus Catastrophe, 1445 B.C.

In the 1400's B.C.E., Moses described, from his Egyptian vantage point, that the land of Israel was a land of "milk and honey." But he had never been there so far as is known, although Moses was well read. Today, San Diego has a latitude of 32° 30 minutes, much like the latitude and the climate of coastal Israel in this era.

Perhaps in 1450 B.C.E., Jerusalem had a latitude of 43 to 45°, like in this age, Eugene and Salem, Oregon, in the verdant Willamette Valley. In the Willamette Valley today, westerly wind systems bring an abundance of rain in the winter and spring. In Western Oregon, there is an abundance of green grass, of cattle, of milk, of flowers, bees and of honey.

But it is not so in arid, sunny Southern California. It is suspected that the description by Moses of the land of Israel, and its abundance of milk and honey, was a literal description for his age, and not figurative one, AS IS USUALLY ASSUMED.

Today, Israel is far too south and far too dry for an abundance of dairy cattle and springtime grass and flowers. Moses' description of a verdant land of Israel seems to be one more indication of spin axis shifts during the Catastrophic Era. Snows reported in Arabia in the Book of Job suggest Arabia also formerly was farther north.

Table X provides data on the orientation of the foundations of thirty ancient Mediterranean Region temples. Wright's three temples at Shechem 11, Shechem 2 and Alalakh in Northern Syria tell a similar story. Why were the location of the spin axis, and the cardinal directions not settled?

The reason is that the Earth experienced a long series of Mars flybys. Each flyby did something to the spin axis. October flybys produced opposite changes compared to March flybys. All in all, this situation left ancient engineers and architects a bit confused. What would be the next appropriate orientation for the structures they designed? Upon what could they depend? The only settled thing was the positions of the fixed stars, their constellations and the 30-degree sectors therein, in short, the ancient zodiac.

The Biblical accounts of catastrophism, interleaved with chronicles, supports alternating torque's for the spin axis of both Mars and the Earth. Kazmann's list of the orientation of the foundations of ancient, Mediterranean region temples suggest the same thing. The spin axis was not secure in that era of Mars-Earth Wars.

Ancient Traditions Of Catastrophism Of The Indians Of North America

There is yet another category of research, although research in depth has not been undertaken. That category is the fabric of stories of ancient times that is preserved in Indian traditions of tribes from across North America.

The Innuit (Alaskan Eskimos) have an account when once a living was easily made by their distant ancestors. Then the earth quaked. Volcanoes erupted. And the path of the Sun in the sky suddenly went low. A cold, harsh climate insured that a living became difficult to achieve.

There is an organization, AISES, the American Indian Society for Engineering and Science (Boulder, Colo.), which addresses the need for American Indian elders to preserve their various ancient traditions from all over the North American continent, from Alaska to Mexico, and from Labrador to California.

Indian elders have found it difficult to preserve their ancient traditions, and in many cases their ancient language, and pass them on to later generations. Then comes the reservation's anthropologists with no concept of ancient catastrophism, and no respect for Indian traditions. They end up competing with, rather than supporting the Indian elders. Thus, on the reservations, young, elitist, white intellectuals have created needless friction between the palefaces and Indian elders, trying to maintain ancient traditions and lore.

Based on data presented in the 1995 AISES convention, Boulder, Colo., it is clear that numerous Indian tribes have heritages of ancient, catastrophic times when the path of the Sun in the sky changed, when they lived longer, and when they fought larger animals.

PREDICTION VIII. It is predicted that when research is completed among the Indian tribes of North America, and collated, no less than thirty different Indian tribes will be found with stories of ancient catastrophic ages, interrupted by catastrophes featuring a change in the Sun's path across the sky.

This all includes sudden climatic changes and sudden latitude shifts. In Volume III, data is cited suggesting that during the closest of all of the Mars flybys, the North Pole shifted 3,000 miles from the North Pacific to the Arctic Basin. Particularly rich will be data in the form of ancient traditions of the Innuit (Alaskan Eskimos), the Mayans, the Hopi, the Algonquins and the Sioux.

Figure 20 - Alternating Torque's of Spin Axis Mars and Earth

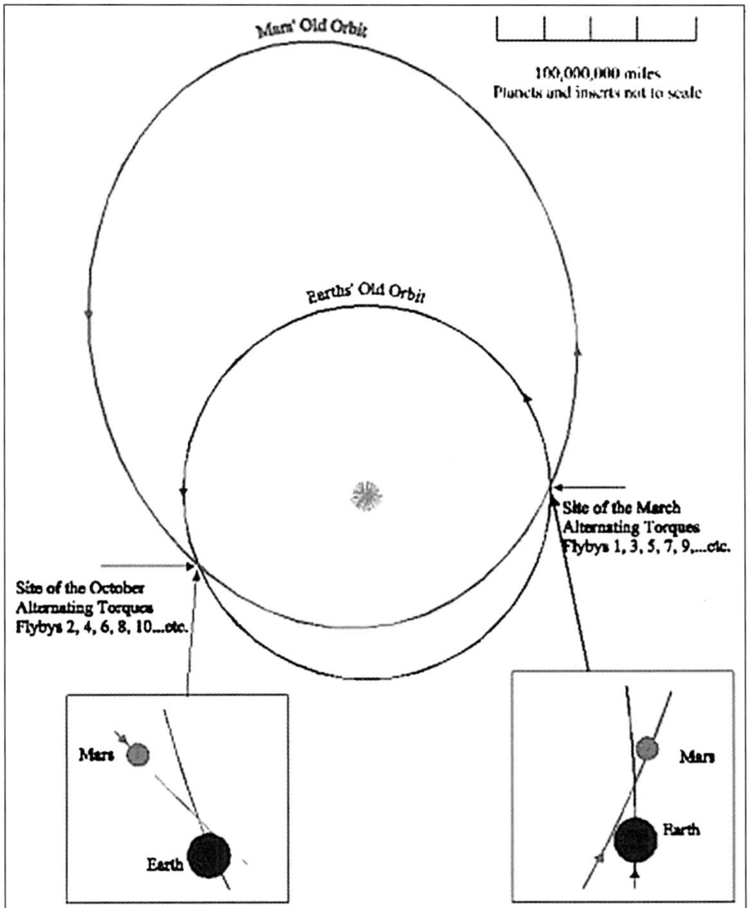

Twin (Alternating) Torques of Spin Axis
Mars and Earth

Figure 21 - Twin Tilts and Twin (Alternating) Torque's of Spin Axis Mars and Earth

A zero degree angle for a torque is the most restful angle. For alternating torques 22.5° is the most restful angle and 45° is the least restful (least stable).

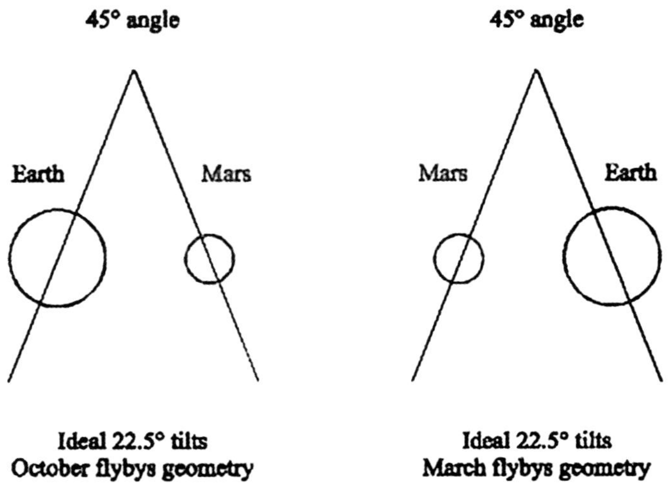

Twin Tilts and Twin (Alternating) Torques of Spin Axis Mars and Earth

The ancient Indian traditions span 5,000 years and go back to what the Greeks called "the golden age." Our data concerning thirty foundations of ancient Mediterranean temples in this chapter spans only the last 750 years of the Catastrophic Era.

The Third Orbit Of Mars And Alternating Torque's

The preceding Figures 20 and 21 illustrate what happened to the Earth's spin axis in a long series of Mars flybys. Figure 23 in chapter 9, illustrates what happened to the Earth's orbit during only the Final Flyby.

It is concluded that on this occasion, the Earth's final waltz with Mars, Jerusalem shifted south by about 4° of latitude. There is 68 miles to a degree of latitude. Jerusalem shifted closer to a new equator, and farther from the North Pole, by about 275 miles.

When gyroscopes receive repeated torques on one side, the most restful angle for the gyro's spin axis is 90° to the torque's, perpendicular. The LEAST RESTFUL ANGLE is neither 0° nor 180°. The LEAST RESTFUL ANGLE for gyros happens to be 45°.

Mars and the Earth experienced alternating torques on the opposite sides of their orbits. They were in March and October in 54-year intervals. See Figure 21. UNDER ALTERNATING TORQUE CONDITIONS, THE MOST RESTFUL COMPROMISE IS 22.5°.

The Earth's spin axis tilt leans at 23.44° to the orbit plane. The spin axis tilt of Mars is a very similar, tilting at 23.98°. They are "twin tilts." Both tilts are within 1.5° of the ideal compromise of 22.5° for alternating torques. This pair of twin tilts for gyros is logical under, and only under alternating torque conditions.

The twin tilts of Mars and the Earth have been known for 200 years, since the era of Herschel. Never has an explanation for their similar spin axis tilts been made in the literature of astronomy to our knowledge. This is the first explanation for this phenomena in the history of astronomy. Many have wondered why the Earth's tilt is 23.5°, but answers have not been forthcoming, until now.

In the gradualist paradigm, an explanation of this twin phenomenon is impossible. In the planetary catastrophist paradigm, the explanation of "alternating torques" addresses all of the data. This explanation is logical, it is consistent with the data, it is consistent with engineering principles, and it is consistent with celestial mechanics.

THE DURATION OF THE MODERN ORBITS FOR MARS, VENUS AND THE EARTH. As of March 20, 1995, the modern era is 2,699 years old. The count is as follows. (a) From 701 B.C.E. to 1 B.C.E. = 700 years. (b) Add four years which medieval monks overlooked, when sequencing the years of history. This error is at the root of the humorous anachronism that Christ was born in the year 4 B.C. (c) Add 1,995 for from 1 AD to 1996. 700 + 4 + 1995 = 2699.

This time span of 2,699 years is compared with the time span gradualist astronomers cite for the Earth being in its present 365.256-day orbit, at its present tilt. Four billion years? 4.6 billion years??? 11 billion years of gradualism? The span of time since the Earth's orbit last shifted is .000000675 of four billion years. One of these two paradigms is a mistake, and a bad mistake it is. .000000675 is also a reasonable assessment as to how close the theory of gradualism is to the theory of planetary catastrophism. Not close.

Every traditional astronomer believes that Mars has never been closer than 30,000,000 miles to the Earth. When Volume III is completed, The Flood of Noah, this planetary catastrophic analysis will be that Mars has never been closer to the Earth than 14,500 miles (planet center to center.)

JAMES CLERK-MAXWELL. It is appropriate to end this chapter with the same words that James Clerk-Maxwell read one evening, 130 years ago. Clerk-Maxwell had been wondering for years what kind of an organization atoms might possess. He was deeply puzzled. That evening he read Job chapter 38.

Just hours later, as he slept and as his sub-conscious took over, he had an inspired dream that the structure of the atom, which nobody understood, resembled the structure of a Solar System. The Sun was a nucleus; electrons were the planets. With that dream coupled with Clerk-Maxwell's reputation and his expertise in physics, science was prepared to take another leap forward. Here are the words Clerk-Maxwell read that evening.

> The Lord answered Job out of the whirlwind, and said, Who is this that darkeneth counsel by words without knowledge?
> Gird up now thy loins like a man; for I will demand of thee, and answer thou me.
> Where wast thou when I laid the foundations of the earth? declare, if thou hast understanding.
> Who hat laid the measures thereof, if thou knowest? or who hat stretched the line upon it?
> Whereupon are the foundations thereof fastened? or who laid the corner stone thereof;
> WHEN THE MORNING STARS SANG TOGETHER, AND ALL THE SONS OF GOD SHOUTED FOR JOY. Job 38:l-7

Conclusion

The purpose of Chapter 8 is to lay the basis for understanding the old orbits of Mars and the Earth, in the Catastrophic Era. This model portrays a long series of alternating torques, Mars-Earth skirmishes, some of which were all out wars.

With gyroscopes, the perpendicular angle to the torque, or series of torques, is the most restful angle. The most chaotic angle is 45° (not 90°.) With alternating torques, the most restful angle is a compromise, 22.5°.

Mars suffered two heavy flybys in the year 701 B.C.E. The foundation for this statement is in Chapters 9 to 11. One was in January, 701 B.C.E., the Last Polka of Mars and Venus. It was followed some 55 or 56 days later by the Last Waltz of Mars and the Earth.

Considering this catastrophic close to the Catastrophic Era, Mars has done exceptionally well. Despite the last Polka and the Last Waltz, it still has a spin axis tilt of 23.98°. This is still close to the most restful compromise, the 22.5° angle.

The Earth's spin axis has done even better; after the last furious flyby in 701 B.C.E., its axis is 23.44°. It is within 1° of the most restful ideal angle.

Story 20 is HOW THE EARTH AND MARS CAME TO HAVE TWIN TILTS. Both tilts approximate the most restful angle for gyros with alternating torques. The only model that explains the twin tilts of the spin axes of Mars and the Earth is the model in this chapter of this book. In the past 200 years of astronomy since the tilt of Mars was first ascertained, to our knowledge no other explanation for the twin tilt phenomenon has ever been offered.

Planets, which rotate in space, respond like gyros to external torques. Mars, Venus and the Earth are fluid gyros, with thin solid crusts. But mostly, these gyros are composed of fluids. The molecules of fluid gyros respond very differently than do the molecules of solid gyros. Soft, fluid gyros have yet to be made projects of scientific research.

Story 21 is that changes in the spin axes of both Mars and the Earth after ancient flybys merely reflect THE NEW AVERAGE PLANE OF THE MOTION OF THEIR MOLECULES.

Story 22 is that magmas mixed broadly under the Earth's crust in flyby scenes. If the materials were silanes, carbides and water, the results were (1) MORE SILICA, OR CRUST, AND (2) THE BYPRODUCT, MORE POOLS OF PETROLEUM.

The silicon and oxygen ions teamed up, the first team, leaving the carbon and hydrogen ions as the second team. They polymerized under conditions of high temperatures and pressures, and formed hydrocarbons, which pooled. The petroleum generally pooled at the top of the mantle.

Much petroleum still exists; it is not caused by dinosaurs dying or vegetation rotting. It was a chemical reaction in the mixing magma during Mars flybys. Much is there; it is just rather difficult to get at for modern industry. The Earth is not running out of petroleum; it is just running out of known petroleum pools that are readily accessible for pumping.

An alcove on story 22 is that the carbon in the atmosphere of Venus likely has a similar origin as does the petroleum on our planet. It too was formed by the mixing of silanes, water and carbides.

Story 23 is that the Final Flyby caused A LATITUDE SLIPPAGE, SOUTHWARD FOR JERUSALEM, OF APPROXIMATELY 4°. This conclusion is based on Isaiah's account.

Story 24 is that the composite spin axis shifts and North Pole relocations OF LATITUDE FOR THE PERIOD 965 B.C.E. THROUGH 701 B.C.E., WAS SIX°, A NET SLIPPAGE SOUTH FOR JERUSALEM. The latitude of the holy city shifted from 38° to its present

31° 45 minutes. This is based, in part, on the 6-degree offset of due east for the foundation of the First Temple (Solomon's) compared to modern cardinal directions.

In comparison, the orientation of the foundation of the Second Temple, (Herod's) agrees with modern cardinal directions. It was laid in the first century B.C.E., in the modern era, and on top of the earlier temple foundation.

Story 25 is that THE SHADOW OF THE SUN DIAL OF AHAZ BECAME SHORTER BY ABOUT 105 INCHES OR 8.75 FEET in Isaiah's account, Isaiah chapter 38. This shadow shift was the equivalent of 4° of latitude for Jerusalem on this occasion. The direction for Jerusalem was a shift southward.

With story 25, the reader is 60% of the way to the penthouse of planetary catastrophism.

Ancient literatures and architectures indicate an abundance of ancient spin axis shifts. The score now has become ancient writing catastrophists 3, modern 20th century gradualists, 0.

CHAPTER 9

The Energy Exchange

... with such tumult Ares the destructive, burden of chariots, charged crying aloud on Herakles who came eagerly to meet him. But now Athene, daughter of Zeus of the aegis, came to stand in the path of Ares, herself wearing the gloomy aegis.

Hesiod , The Shield of Herakles, Lines 441-444

I should like to recommend a remedy to you. I refrain from the effort of trying to establish a new theory for Mars. For it is already established. I have spent so much pains on it that I could have died ten times. But satisfied with my discovery and rest assured, out of my peace you might draw an end for these discoveries.

Johannes Kepler

Johannes Kepler (1571 - 1630)

So wrote Johannes Kepler in 1605 to his friend Fabricius. The letter concerned the elliptical nature of the orbit of Mars. By this time, Kepler was a confirmed Copernican. And he was in Prague, an associate of the famous Danish astronomer Tycho Brahe. Tycho was the best star watcher and recorder of star positions of that time. When Tycho died, Kepler inherited his voluminous records of star and planet positions. This was a big assistance in his work.

Above is part of a letter from Kepler to David Fabricius, who was an independent-minded thinker, but who yet kept contradicting Kepler on his discovery. Like Copernicus, Fabricius preferred that Mars have a circular orbit, not an elliptical one, and that the red planet's velocity in space never changed. Kepler knew better.

Kepler had carefully studied the orbit of Mars and had logged the positions of Mars in its present orbit. From his studies, and supported from Tycho's records, a second opinion, Kepler concluded that Mars advanced more slowly in one part of its orbit, near its aphelion, and faster when nearing its perihelion.

Kepler had calculated and determined that the Martian orbit was an ellipse; it was not a circular orbit as Copernicus and Fabricius had assumed. This was the start of his scientific breakthrough.

Kepler was well on his way to codifying his famous three laws of planetary motions. This codification, published in 1626 twelve years later, was his biggest achievement - and the most painful achievement to the more numerous astronomers, flat Earth folk, who preferred Ptolemy, and who preferred having the Sun to orbit the Earth.

In producing his three laws of planetary motion in 1626, his Rudolphine Tables, Kepler became the father of celestial mechanics and of modern astronomy. This publication in 1626 became one the strongest of a long series of birth pangs that birthed modern science. [n1], [n2]

It is to be observed that Kepler's work featured a study of the orbit of Mars in the modern era. That orbit is now known to have an eccentricity of .0933865, a perihelion of 128,409,085 miles and an aphelion of 154,862,861 miles. Its perihelion has a longitude of 335°, which places its semi-major axis in space. Its modern period is 686.978839+ days, a period which began in 701 B.C.E.

This present research is a study, including calculations, of the likely orbit of Mars in the Catastrophic Era. Shortly, the energy calculations and the shifts in energy for the ancient orbit of Mars will be presented. In the next chapter, the simultaneous angular momentum shifts and their calculations are presented. Both address the heart of celestial mechanics.

The key date is March 20-21, 701 B.C.E. This was the night of the Earth's final waltz with Mars. This occasion as recorded by Isaiah was the 15th year of the reign of King Hezekiah. Edwin R. Thiele has produced a classic in dating the era of the kings of Judah and Israel. In his historical analysis, The Mysterious Numbers of the Hebrew Kings, Thiele concludes the year of the key flyby, Hezekiah's 15th, was 701 B.C.E. [n3]

Our astronomical study, based on retro-calculations of the orbits of Mars, Jupiter and Saturn, independently places Mars there at this time. Retro-calculations for 701 B.C.E. also place Jupiter in the feared and significant Capricorn, and Saturn at 180° opposite, in the feared and significant zone of Cancer.

Introduction

If planetary catastrophists were unable to demonstrate logically - and could not model - how Mars shifted from its "catastrophic orbit" to its modern orbit, reservations and indeed full-blown disbelief about planetary catastrophism would be appropriate.

In that era, this analysis models the orbital period of Mars as having been in following resonance's:

With the Moon 1:24
With Jupiter 6:1
With the Earth 1:2
With Uranus 85:2
With Venus 5:16

In addition, there was a near perfect ratio of the orbits of Mars and Saturn, very close to 15:1.

If, on the other hand, a model can be developed wherein energy shifts and angular momentum realignments are solved simultaneously, and work in harmony, then it can be called a "scientific model." Then, tit for tat, reservations and full disbelief about the gradualist system would be appropriate. Models are two-edged swords.

This study is about the Catastrophic Era orbit of Mars and the change of that orbit. Its orbital eccentricity was approximately .561; its perihelion was about 64,353,000 miles and its aphelion about 228,805,000 miles. Its longitude of perihelion was 105° and its period was 723.26 days (modern days, not Catastrophic Era days).

It was an age of orbital "harmony in the heavens," an ancient phrase used notably in the Book of Job. More appropriately, the former age was an age of orbital harmonics. All this ended in the year 701 B.C.E.

How Did Mars Achieve Its Modern Orbit?

If the shift in the orbit of old Mars to modern Mars, herein advocated, cannot be demonstrated, any and all discussions of Mars-Earth catastrophism are cloudy and uncertain.

Therefore for science, the next four chapters, 9, 10, 11 and 12, are the most important chapters in this volume. The heart of these chapters are Tables XI, XII and XIII, what are named our "Rudolphine Tables". In chapters 11 and 12, twelve clues of scientific support are presented. The aim is to present a triad of truth - energy shifts, angular momentum shifts and solid supporting clues.

Tables XI and XII of the Rudolphine Tables are condensations. Table XIII, the most detailed, addresses those ancient orbital shifts in 24 categories of astronomical data. This data defines precisely what happened.

It is incumbent for serious students of science and cosmology to delve into this data --- statistics of orbits and of planet masses that comprise planet energies and planet angular momentum's for four planets involved.

For those of non-mathematical backgrounds, the significance of the ensuing data may be lost in "boring" numbers. In such an event, skipping over the upcoming statistical material in favor of the more descriptive material is appropriate. But it is with the recommendation that later, the Tables XI, XII and XIII be reexamined.

It might be what all gradualists assume, that Mars has been only a slightly reddish pinpoint of light in the nocturnal heavens since the beginning of time. It might be that Mars never has come closer to our planet than 33,900,000 miles. Moreover, it might be that Mars never orbited out further, into the asteroid belt, some 230,000,000 miles distant from the Sun. If so, why are 93% of its craters on one side, and how did it capture its little satellites on the fly?

If that were the case, this volume and even this series might have to be classified with such books as Alice in Wonderland, The Wizard of Oz and the Star Trek themes. Fantasy. Historical fantasy and science fiction.

However the converse also is true. If this model can be defended successfully in celestial mechanics, it logically follows that some version of the data herein is solid history --- solid history of the solar system -- solid cosmology. Involved are the changes to orbits for Venus, Mars, the Earth-Moon system, Jupiter and ... Astra.

Like the dogma of Claudius Ptolemy, the 18th-19th-20th century theories and dogmas of gradualism would then become recognized as fantasy, imaginative and just so much science fiction. There is no mistaking it; at least one of these two competing paradigms of solar system history is applesauce.

Fifty years ago, virtually all academic astronomers and geologists classified themselves as millions of years "gradualists." However times do change. Photographic evidence has come in showing heavy catastrophism of the surfaces of Mercury, Venus, the Moon, Mars, Deimos, Phobos, Io, Europa, Ganymede, Callisto, Titan, the Rings of Saturn, and other satellites more distant.

In general, these surfaces reveal surprising spasms of catastrophism, whatever may be the era. The evidence dictates that some events there were rather energetic.

Due perhaps to the planetary missions of the last 30 years, now, in the 1990's, almost all astronomers and geologists prefer to be known as "Millions of Years Catastrophists." "Catastrophism" has become the latest vogue, an acceptable label of academic fashion.

This is a 180-degree change in the preference of styles over the last 40 years since the early first space probes. In the 1940's and 1950's, catastrophism had about as much respect as prostitution. But the changes that have occurred may have been one of style, not one of substance, not a change in paradigm.

Judging from the scenario of planetary catastrophism, many "lately styled catastrophists" actually cling to most of the traditional assumptions and dogmas of gradualism, whether or not they are aware of it. There hasn't been a competing paradigm, except perhaps among the followers of Immanuel Velikovsky. With the model herein, the dogma of "Millions of Years For Catastrophism" has competition. Not only are the nature of catastrophic processes at issue; so are the timing issues.

In geology, like astronomy, the new style of the 1980's and 1990's also has become catastrophism. There are paleo-catastrophists, meso-catastrophists and ceno-catastrophists in addition to astronomy, where there are "asteroid catastrophists" and "comet catastrophists". To repeat, in 1996 hyphenated catastrophism is in vogue, but no basic paradigms have changed. Until paradigms shift, the change toward "catastrophism" is more one of style than of substance.

Some still assume what has been so widely taught, that the Sun had a hiccup or two, and a cough or two, for reasons unknown, four billion years ago. Expelled out was the makings of a planet or two, or three or four, or six or eight, even nine. Yet the planets we encounter have spin rates, often in pairs, craters, satellite systems, etc., all unexplained. Such is the dysfunctional, traditional menu of gradualism.

There is evidence that planetary catastrophism has assaulted Venus, the Earth, Mars and has destroyed Astra. In addition, it is pointed out in our "The Recent Organization of the Solar System", there are four levels of evidence that planetary catastrophism has also extended the heart of this solar system, to the Sun itself.

The only conclusion we can draw from the evidence is that the Sun itself had a nova recently, as astronomers chart time. The Sun is still cooling and shrinking, recovering adjustments of that crisis to the Sun.

As to asteroid catastrophism, asteroids typically are .0000000002 of the mass of the Earth; about one twentieth of a billionth of the mass of the Earth. Such masses can create surface craters of various sizes.

But a single asteroid collision with the Earth cannot create a spin axis shift, an orbital shift, a geomagnetic field dynamo (or generator), a fragmentation (icy or rocky), or a paleomagnetic polarity reversal, much less sudden crustal tears and up thrusts, a hemispheric sized flash flood and sudden sedimentary strata. Our planet, the Earth, has experienced all of these. Evidence indicates the timing thereof has been "recent".

Comets, typically are icier and smaller than asteroids. Typically, they are tiny, dirty snowballs with orbits of high eccentricity. Were a small ice ball to enter the Earth's atmosphere, like the Tunguska Bolide on Northern Siberia, there would be considerable damage on a local scale. The Tunguska Bolide, June 30, 1908, 7:17 a.m., was just such an event.

It created a flash explosion, with a shock wave heard for over 1,000 miles. It decimated forests, blasting them down and burning them up, all within a 10-mile radius. But a colliding comet could never cause (1) a spin axis shift of several degrees, (2) an orbital relocation, (3) a paleomagnetic polarity reversal, (4) a hemisphere-wide flood or several other phenomena associated with ancient planetary catastrophes. Never. These scales are entirely different.

In geology, all newly styled catastrophists have to decide if they are paleo-catastrophists, meso-catastrophists or ceno-catastrophists. But they have yet to identify the causing agent or the scope or the timing of these catastrophes. They have yet to consider Mars making a planetary flyby as close as 100,000 miles or (heaven forbid) 27,000 miles to our planet.

Thus, planetary catastrophism is neither asteroid catastrophism, nor is it comet catastrophism. Not in agent of cause, not in scope, not in timing. The energy exchanges in planetary catastrophism far exceed such minuscule explanations. And the timing of these upheavals is much too recent.

The Challenge

To reiterate, if planetary catastrophists cannot demonstrate logically - if it cannot be modeled - how Mars shifted from its "catastrophic orbit" to its modern orbit, then this model of planetary catastrophism is in trouble. Also, conversely, if it can be sensibly modeled, 18th century gradualistic cosmology is in trouble.

Chapters 9, 10, 11 and 12 are dedicated to this task, and this achievement. For many general readers, who are comfortable with popularized materials, the discussion will now move into an area with which they are unfamiliar, celestial mechanics. However, to accent the positive, mind-bending material can be good; it can open up new vistas of thought and new perspectives of events in ancient history.

This is the first of four chapters that address how and why the orbit of Mars shifted from its Catastrophic Third Orbit to its Modern Era Orbit. This chapter addresses energy shifts. Chapter 10 addresses angular momentum shifts between Mars and Venus, the Earth-Moon system, and Jupiter.

It is found that the two energy shifts were less than 60 days apart. The last of the Mars-Venus polkas briefly preceded (by 58 days) the last of the Mars-Earth waltzes.

Chapters 11 and 12 concern clues, twelve in number. In murder mysteries, clues are left at the scene and sometimes elsewhere. In the death of the Catastrophic Third Orbit of Mars, twelve clues exist, and they can be sifted from the ancient "wreckage." Each clue points to the crime, its timing, and to its perpetrator, the red planet.

Many physical scars on the surfaces of Mars, the Earth and the asteroids have been discussed in Chapters 1 to 8. More are discussed in Chapters 9 to 11. In a change of direction, Chapter 12 discusses psychological scars that still remain with us, 2,700 years later. It discusses over 350 of them.

Energy and Angular Momentum

ENERGY. In a closed system, energy can be given up by one body to another and it can change its form as well. One planetary body can acquire (or lose) a specific amount of energy. But at the same time, another planet must give up (or acquire) an equal amount. Such is the second law of thermodynamics. Such is the principle of the conservation of energy. Nature knows no exception in a closed system, such as our solar system.

The total amount of energy in a closed system must always total the same - at the beginning, at the end of each stage in between, and at the end of an entire series of stages. This process is what happened to four planets, rapidly, late in the 8th century B.C.E.

Because no "exchange" is perfectly efficient, some energy always is transformed into heat, tides, etc. This is insignificant both in billiards and in planets where calculations are taken to six decimals.

ANGULAR MOMENTUM. Similarly, in a closed system, angular momentum also can shift from one Solar System body to another nearby planet. But at the same time, what one body gains in angular momentum, another planet must lose, and this is also true at the end of each and every stage. Again, as with energy changes, nature knows no exceptions.

Perhaps one can visualize the Solar System, looking down on it from above, viewing it from the North Star. In this format for a planet in orbit, one can visualize ENERGY as "vertical motion". This is a measure of its long axis multiplied by planet mass.

This is the "x" axis of its orbit. Energy changes relate to any and all vertical changes along this axis. Changes can be in length, or in direction or both. Energy is a total property of a body that relates to its mass, its speed, its temperature, etc., but not to its direction of travel.

In this same context, one can view ANGULAR MOMENTUM as "horizontal motion," or sideways movement. Its changes affect only an orbit's "y" axis, the measure of an orbit's widest dimension. It is called the minor axis. Momentum relates not only to a planet's mass and its velocity but also to its direction. To the mathematician, energy is a scalar value, while angular momentum is in vectors.

There is a similarity between shifts - (a) of energy and (b) of angular momentum among the planets. It is (c) the completion of a crossword puzzle. Crossword puzzles are composed of both horizontal words and vertical words. To solve it correctly, all the letters in vertical words must agree with all letters in the horizontal words. Otherwise, the solution for the puzzle is flawed.

Energy can be likened to the vertical words of a crossword puzzle. Angular momentum can be likened to words on the horizontal axis. In a correct solution, all words and letters must agree, harmonize and integrate.

If both of these two issues are found to be in agreement, simultaneously, it can be concluded that the puzzle is solved (or nearly solved). If one issue is not addressed, or is not in agreement, the puzzle is not solved. Like crossword puzzles, there is one totally correct solution, and only one.

Put another way, if the two issues above are found to be in agreement, it can be said that it is a completed model that conforms to the Second Law of Thermodynamics. It is a scientific model. It can be verified by known data, equations, procedures and testing.

On the other hand, if a model of catastrophism does not conform to these laws, it can be said that such a model is imaginary, wishful thinking, and/or speculative. It could even be the stuff from which fantasies, wild speculations, fairy tales, false superstitions and science fiction plots are spun.

The Circulization Of The Orbit Of Mars

Issue 1. Mars And Astra Distances

There are five prominent issues involved in this model, and in understanding the shift of Mars from its catastrophic orbit into its modern orbit. The first issue concerns whether or not it was the Roche Limit of Mars - and hence the mass of Mars - that caused little Astra to fragment into asteroids.

If this was the case, then when Astra fragmented, Mars must have been some 225,000,000 miles distant from the Sun, where the average perihelion's of the asteroids is. This is contrary to traditional theory. Mars nevertheless was there. The red planet was 70,000,000 miles farther out, or more, than is the modern orbit of Mars. Probably, its ancient aphelion was even somewhat farther out.

If Mars indeed was there, its energy level would have to be maintained within a few per cent. The center of its orbit would be not far different than now. With this in mind, if Mars went out 70 or 80,000,000 miles farther, also it would had to have come in much closer. Its orbital length and its eccentricity would have to have been greater.

In so doing, Mars would approach closer to the Sun than the Earth, and even closer than

Venus in a long narrow orbit. The central issue then is identified as the old eccentricity of the orbit of Mars. Its modern eccentricity is .093387. A prefect circle has an eccentricity of .000000. THE CENTRAL ISSUE THEN IS WHETHER OR NOT THE ORBIT OF MARS EXPERIENCED A SUDDEN AND SUBSTANTIAL SHIFT IN ITS FORMER ECCENTRICITY.

Its ancient orbital eccentricity would had to have been greater than .400000 in order for its perihelion to be within the orbit space of the Earth-Moon system. And even more, its ancient orbit eccentricity would had to have been over .500000 in order to orbit within Venus orbit space. Its modern eccentricity is under .10, at .093.

Therefore research for this model requires a sudden change in an ancient eccentricity from the old Mars orbit in excess of .400. The issue is not one of a vast shift in the total orbit of Mars. It is much less; it is a vast shift only in its eccentricity, or oblateness.

This model has the ancient orbit of Mars, from perihelion to aphelion, with a length of 293,158,820 miles. Its modern orbit from perihelion to aphelion is 283,271,946 miles. In this model, the length of the semi-major axis of Mars has diminished, but only 3.37%. Mars lost some energy and in so doing, the length of its "x" axis was reduced by 9,886,874 miles. The Martian energy diminished from -1.345017 to -1.39196l.

What was the cause of such a reduction? Did another planet's energy similarly increase? Was that planet the Earth? Did ancient calendars of 360 days suddenly become obsolete, requiring a higher day count per year, reflecting an increase in energy? An increase from 360 to 365.256 is an increase of about 1.5%. What is this story?

This staff also has calculated that the probable former orbital eccentricity of Mars was .561. Its orbit was lower in eccentricity than Halley's Comet, but nevertheless, it had a high eccentricity. This compares to the modern eccentricity of the Martian orbit at .093. It is a shift downward of .468.

Its follows that in the catastrophes of the late 8th century B.C.E., the width of the "y" axis, the minor axis of Mars increased. This model indicates the length of its minor axis (the "y" axis) increased by 19,672,856 miles as its orbit rounded out. This is why Mars no longer bothers Mars, and it no longer bothers the Earth-Moon system. It was the increase in angular momentum that rounded out the orbit of Mars, with the very significant results that celestial peace came to the inner solar system.

Thus it is seen that the long axis of the orbit of Mars was shortened, but not a lot. But its short axis, the "y" axis increased by 16.2%. In this process, the period of ancient Mars was reduced from 723.26 new days down to the current 686.98 new days.

Does the shortening of the semi-major axis of Mars integrate well with the evidence that the Earth's ancient period increased from 360 to 365.256 days? Of course. Does an increase in its semi-minor axis and its rounding out agree with the observation that Mars no longer threatens either the Earth or Venus? Again, of course.

The Rudolphine Tables, XI, XII and XIII indicate that the Earth's semi-major axis did increase, from 92,339,242 miles to the modern 92,955,807 miles. It was an increase of 0.668%. One of the results was that all of the old 360-day calendars used by numerous ancient societies suddenly became obsolete. Calendars of 365, 365.25 or 365.25+ days were needed in replacement. Did this happen in the 7th and 6th centuries B.C.E.? Most certainly, according to ancient testimonies from a wide variety of sources. .

Issue 2. The Moon And An Ancient 30-Day Orbit

The second issue concerns the Moon and whether or not some 12 to 15 ancient lunar calendars, each containing 30.00 days for the old lunar orbit, were accurate and hence useful in the ancient era. If it were not the case, why are there so many ancient calendars from diverse peoples and places all making the same mistake? After all, ancient calendar makers were not Neanderthals; they may have been basically more intelligent than modern men.

If the period of the Moon's orbit shifted, some event such as a planetary flyby must have occurred, and affected both the Earth and the Moon simultaneously. It reduced the Moon's period from about 30.00 old days (sidereal) down to 29.53 new days, its modern period.

If ancient calendars contained 30 days, not 29.53 days, the Moon's orbital radius necessarily was decreased from about 241,500 miles to the modern distance, 238,900 miles. If so, this was a shrinking of 1.08% in the radius of the Moon's orbit. Could such a strange change happen if Mars passed between the Earth and the Moon when at full? Yes it could. The Earth would be pulled outward while Mars would be pulled inward.

How could the expansion of the Earth's orbit be related simultaneously to the contraction of the orbits both of the Moon and of Mars? Does this mean that the Earth gained energy or that the red planet lost energy? Yes for both.

Both of these are issues are easily answered IF MARS "BUZZED THE EARTH" FROM THE NIGHT SIDE OF THE EARTH, SOMEWHERE BETWEEN THE EARTH AND THE MOON, that is, DURING A FULL MOON. The ancient Hebrew "Passover" was always recorded to have occurred on the night of the first full moon of Nisan (or our March). The "Passover" of -701 featured just such a full moon.

On their Mosaic calendar, the Passover was Nisan 13, always a Friday night in the Hebrew calendar; it was the "unlucky thirteenth". (It corresponds to the night of March 20 on our modern calendar). This night, once every century or so, actually involved a devastating Mars flyby. Research to be published later indicates repeated Mars flybys in 108-year cycles.

Ancient traditions of a catastrophically unlucky Friday the thirteenth is a vestige from Hebrew traditions when, sometimes, Passovers on that date were very destructive. The Romans have a similar tradition, their tubulustrium, a time of trouble, also March 20, appropriately, in a month they named after Mars.

The Irish, on the other hand, are not to be outdone. There were October 24 flybys also, which were even more destructive to the Eastern Hemisphere than were the March flybys. Our modern Halloween themes are a carry over, a vestige from ancient Celtic traditions about destructions coming visibly from the heavens, in late October, at the time of a full moon. The comet Mars' tail was interpreted as the witch's broomstick on which she flew. It is the Irish interpretation comparable to the Greek "Fleece of Aries".

Occasionally the cosmos of Celtic Ireland was fraught with celestial danger - something witchy, something moving across the heavens, making screechy celestial noises, with an orbit and a cometary tail instead of by broomstick. That something was something close enough and massive enough to produce loud noises from lightning, shock waves, earthquakes, weird subliminal noises at subliminal levels, possibly due to uniting planetary magnetic fields.

In the Hebrew calendar, March 20 or Nisan 13 always was a full moon ... and Friday the 13th - of the month of Nisan. If the Moon always was full on Nisan l3, and the ancient Hebrews had twelve 30-day months, Mars had to be in some kind of orbital resonance with the Earth. It had to be. Therefore it is important to note how many other societies around the Earth also had 360-day calendars (and 360-degree circles). This will be addressed in chapter 12.

In the modern era, lunar calendars don't work well for anniversaries. Passover nights, on full moons, are always on a different day each year, like the dates for Easter. In this age, Passover celebrations and Easter Sundays are linked to the first full moon AFTER the vernal equinox.

In that era, the full moon of March/Nisan always was the vernal equinox. Ancient Hebrew calendars portray this. In this age, 2700 years after the change in orbit occurred, lunar calendars still are employed to determine Easter and Passover, a vestige of ancient 30-day calendars.

Periodically, repeatedly, in 108-year cycles, the Destructive Angel of the Lord arrived with "pestilences", possibly involving geomagnetic waves, accompanied by cosmic lightning discharges, fire from heaven, earthquakes, deforming crustal bulges, volcanic upheavals, oceanic tides, and a scintillating geomagnetic show. Talmudic sources identify the night of the Final Flyby as March 20-21. Our research indicates March 20-21 of the year 701 B.C.E. The word "Passover" is, and was quite appropriate. [n4]

Issue 3. The Earth And An Ancient 360-Day Year

The third issue concerns whether or not the Earth's orbit suddenly expanded from a former 360 day count per year to a modern 365.256 day count. Such a shift would involve a sudden expansion of the day count per year by about 1.003%.

This would necessarily involve an expansion of the average radius of the Earth's orbit from approximately somewhere around 92,339,242 miles to the modern value of 92,955,807 miles. This is an expansion of 616,000+ miles in the radius, 0.668%.

If catastrophism happened as this model describes, that perturbation by Mars was more than a mere tweak on the Earth's orbit. It was a strong yank on our orbit, with simultaneously a strong gyroscopic torque on our planet's spin axis. [n5], [n6]

If this occurred during the last Martian flyby event, it could occur only if Mars passed the Earth's outside, which is as Isaiah and Talmudic source material both report to have been the case. Mars in orbit would have to have lapped the Earth-Moon system, somewhere between the Earth and the Moon. And the moon would have to have been at or near full, as in "a Passover night".

Such an orbit expansion for the Earth and orbit contraction for the Moon could happen, but only if that planetary flyby was between the two, but much closer to the Earth than the Moon.

The Moon's mass is 1.23% of the Earth's mass, about one part in 81.5. Mars is 10.74% of the Earth's Mass. The Moon is 11.45% of the mass of Mars. Thus, roughly, the Earth is nine times the mass of Mars, and Mars is nine times the mass of the Moon.

In order to create these kinds of orbital perturbations, Mars would have to go on the night side of the Earth, during a full moon. And the red planet's distance to the Earth at its closest works best if it was between one-eighth and one-ninth of the Earth's distance to the Moon. This means that Mars, at its closest, was between 26,500 miles and 30,000 miles from the Earth, center to center. And Mars was, at its closest, 215,000 miles from the Moon, again center to center.

Orbit perturbations (directional changes or shifts) caused by one planet on another follow an equation that states changes are according to the mass of the two planets, and are inversely proportional to the distance of the flyby squared.

This means that, during flybys, each time the distance between Mars and the Earth was halved, as from 240,000 to 120,000 miles, orbit shifts and energy shifting, whatever the amount, increased fourfold. Thus, at a flyby of 27,000 miles, the orbit perturbation was four times as much as at 54,000 miles. And perturbations would be sixteen times as much as flybys at 108,000 miles. And 64 times as much as at 216,000 miles.

During a CLOSE, OUTSIDE flyby, Mars could shift its directions --- somewhat --- its orbital direction now could pivot on the Earth. Such a close outside flyby by Mars would produce a push or a torque on the Earth's spin axis, producing a minute increase in the rate of its rotation. Such a close flyby could, and apparently did increase the Earth's spin rate, however minutely.

The model presented in the next three chapters maintains there was an increase in spin rate due to the Final Flyby of Mars, and its torque. That speed up is calculated at approximately 0.452+%. Thus, it is modeled that the spin rate increased from about 360 old days to 361.628 new days, an increase of 1.628 days or 0.452+%.

Simultaneously the orbit expanded. The combination of the new orbit and the new spin rate produced an increase from 360 old days (= 361.628 new days) to 365.256365 days per orbit. Of the combination producing the new increase in day count, this model proposes about 31% was due to the tiny increase in spin rate, and the rest, about 69%, was due to the expansion of the orbit, an expansion over 600,000 miles in the semi-major axis.

Today's spin rate relative to the fixed stars is one rotation in 1,436 minutes. This is "sidereal time". All gradualists suppose that this spin rate has been unchanged in 4.6 billion years. Our model suggests that before 701 B.C.E., the Earth's slightly slower spin rate, was between 1,442 and 1,443 minutes (sidereal). This is an increase in spin rate of 0.452%, or one part in 221.

The ancient increase in day count per year to 365.256 days per orbit is a product of a slight increase in spin, accounting for some 31% of the new day count per year. And there was a significant increase of the Earth's distance from the Sun, accounting for an estimated 69% of the new day count per year, 365.256 days. As the Earth's distance from the Sun increased, there was a corresponding slight decrease in the Earth's velocity around the Sun.

The increase in spin rate seems to have been 0.452%. The increase in orbital radius, the semi-major axis, seems to have been 0.668%. The increase in day count per year, a combination of the two preceding factors, has totaled about 1.46% according to this model. This means 360 old days were equal to 361.628 modern days. The spin rate was a minute bit slower.

This analysis is in harmony with a unique, Romanesque ancient change in the day count per year for the Earth made by a mathematical philosopher, Plutarch. Plutarch said and wrote that at the end of the heroic age, the era of catastrophes, there had been a "celestial crap game". It was between Hermes, the Earth and the Moon. Plutarch, or his translators, apparently confused Hermes with Ares.

In that celestial crap shoot, the Moon lost 1/70th of her "holdings", or its period, while the "winner," the Earth gained a similar 1/70th of its day count per orbit, an addition to its former orbit period.

The Moon's modern period is 29.53 days. 30 days minus one part in 70 is 29.57. Plutarch was within .04 day of being exactly correct for the new lunar period.

The Earth's new period, 365.256 days, less one part in 70, results in an earlier period of 360.038 days. Here again Plutarch's explanation was within .04 of a day of being exactly correct. Plutarch's ancient Greek sources were solid, and his explanation for the new conditions satisfied his Roman audiences.

Issue 4. Mars And An Ancient 720-Day Year

The model here proposes that Mars had an ancient 720-day orbit (or 723.257 new days). Mars was in 1:2 resonance with the Earth, and it was in 6:1 resonance with Jupiter. Further, it was in a near 15:1 resonance with Saturn. The general pattern of perturbations of Mars by the Earth and Jupiter requires this resonance of periods.

Interestingly, James Frazer cited in "The Golden Bough" that the Gonds, a tribe in Southern India, still worships Jupiter and has celebrations for Jove every twelfth year. This relic of ancient planet worship is despite the fact that the Earth has not been in a 12:1 resonance with Jupiter for 2700 years.

A contraction or shrinkage of the orbit of Mars from 720 old days (or 723.26 new days) down to the modern 687 new days per year is a change in the right direction. The Earth gained in day count per orbit - 5.256 old days - while the orbit of Mars, one ninth of the mass of the Earth, lost 36 days. What one planet gained in energy (and in angular momentum), the other lost, a classical exchange of energy.

Some have superficially assumed that this is a three body problem, involving just Mars, the Earth and the Sun. No. It is a five body problem, involving Mars, the Earth, the Sun, Venus and Jupiter.

The model presented below portrays that the spasms of catastrophism in 701 B.C.E. were a celestial double header. First there was an expansion of the Martian orbit which caused a close, inside flyby of Venus. Tables XI and XIII indicate Mars gained from Venus .086476 energy units.

Some 54 days later, the red planet encountered the Earth and lost .133424 energy units to the Earth-Moon system. For Mars in 701 B.C.E., it was a net loss of .046944 energy units. There was also a loss in angular momentum.

The net result for Mars of the two 701 B.C.E. crises was a contraction in the period of Mars. Measured in terms of days, it was 36 days - about 5.016%. In terms of its average distance to the Sun, the two planetary flybys of Mars in 701 B.C.E. resulted in a net contraction in distance of 3.37%. Again see Tables XI and XIII.

If Mars passed close by, on the outside of the Earth, because of the Earth's much greater mass, our planet would (and we affirm did) pivot and pull Mars inward eight or nine times as powerfully as Mars could pull the Earth outward. The pattern is consistent. Ancient calendars indicate something very much like this happened. Ancient folklore's involving Earth upheavals with the planet Mars (by whatever name) occurs on at least five continents.

As was mentioned earlier, Plutarch, a first century AD philosopher taught the rich, gambling, semi-barbarian Romans that the Earth and the Moon once had entered into a celestial crap game with Hermes. Actually Plutarch made a misidentification; it was with Ares, not Hermes (Mercury.) Plutarch taught that the Earth gained one-seventieth of its period in that celestial crap game. 365.256 x 69 / 70 = 360.038 old days. That was good enough for the "undereducated" (barbarian) Romans.

Plutarch also taught that in the same celestial crap game, the Moon lost one-seventieth of its period. 30 days x 69 / 70 = 29.5714. The Moon's modern period is 29.5306 days - its "lunation" or synodic month. Again, Plutarch's explanation was good enough to satisfy Roman standards about both celestial mathematics and gambling. [n7]

Energy And Its Equation

The equation of energy for any orbiting planet involves four factors. Two factors are constants, two are not. The first constant is (a) two Pi squared, which is 19.739209. The second constant is (b) a multiplying factor needed to express a planet's energy in usable measures.

Of the two variable factors, the third is (c) the masses of the two celestial bodies involved in the perturbation. The masses of the relevant planets are as follows:

Venus	0.814988
Earth-Moon System	1.012303 (Earth = 1.000000)
Mars	0.107447
Jupiter	317.799873

Venus	0.723332 a.u.
Earth-Moon System	1.000000 a.u.
Mars	1.523692 a.u.
Jupiter	5.202804 a.u.

The fourth factor is orbital. It is (d) the length of the "semi-major axis" of these four planets, formerly and in the current age. The semi-major axis is half of the distance of the long axis, the "x" axis. It also is half of the distance from either aphelion or from perihelion to the midpoint of the orbit. The semi-major axis of the relevant planets are given in "a.u." One a.u. is 92,955,807 miles, the average distance between the Sun and the Earth.

The energy lost by the Moon, involved in the shrinking of its orbit some 3,000 miles, can also be calculated. But the calculations involved require going up to twelve decimal places. To employ such figures for the Moon's energy and the contraction in its orbit would give a false impression of extreme accuracy. Therefore, for both energy balances and for angular momentum considerations, changes in the length of the axis of the Moon's orbit are so slight that they are dismissed.

On the other hand, the mass of the Moon is .012303 of the Earth, about one part in eighty-one. Earth's mass is 1.000000 Earth masses, the measure of one unit of weight. Together, the mass of the Earth-Moon system is 1.012303. The addition of the lunar mass to the Earth's mass is significant and necessary for purposes of calculating perturbations for both Mars and the Earth.

Issue 5. Astronomical Support Data

There are several statistical categories of astronomical data that support the model. One can retro-calculate the position of Venus to determine whether or not it was where our model requires it was on January 24, 701 B.C.E. Retro-calculations are made on the orbits of the Moon, Venus, Mars and Jupiter for their locations on Jan. 24, 701 B.C.E. for Venus, and March 20 for the Moon and Mars on the night of March 20, 701 B.C.E. See Chapter 11.

If they retro-calculate well, they will comprise four clues of support, evidence to indict Mars as the ancient "bane of mortals". If retro-calculations are unsuccessful, they will invalidate this model.

Retro-calculations can be made from any modern date, back to March 20, 701 B.C.E. Were the Moon, Venus and Mars where our model claims they must have been? Was there a Mars-Venus interaction about January 24, just before the Final Flyby? Can the orbit of Mars, and the position of Mars therein, be traced back to March 20, 701 B.C.E.?

For technical reasons, the energy of a planet is always cited in negative numbers. Thus, the higher the negative number, the LOWER the energy level, and vice versa. The lower the number, the greater the planet's energy. Negative numbers for describing energy are helpful in science but make a confusion for the regular non-scientific reader.

Among Kepler's works are Mysterium Cosmographicum (1596), Astronomia Nova (1609) and Harmonice Mundi (1619). Kepler is considered the founder of physical astronomy because of his demonstration that the planes of all planetary orbits pass through the center of the Sun. He recognized the Sun as the moving power of the planetary system.

In his first teaching position at Graz, Johannes Kepler undertook extra-curricular nocturnal studies of the successive positions, hence, of the movements of Mars. Years later in Prague, this Mars data taken 400 years ago led to his monumental scientific breakthrough.

All planetary orbits are ellipses that follow three laws of planetary motion. Kepler discovered those laws and published them. Two are in his original Rudolphine Tables. Emperor Rudolph of Prague was his benefactor, and was the one to whom he dedicated them. It is to Johannes Kepler, that our Tables XI, XII and XIII are dedicated; they are our "Rudolphine Tables."

Table XI is a resume on energy changes. Note that Jupiter energy did not change very much. Table XII in the next chapter is a resume on angular momentum changes. There, Jupiter's angular momentum does shift a bit. Table XIII, also in the next chapter, is an in-depth analysis. Table XIII lists 24 astronomical categories, involving the four planets and the various measurements of change from the Catastrophic Third Orbit of Mars to the Serene Fourth Orbit. These shifts were what produced the modern orbit of Mars.

Table XI - The Energy Exchange of 701 B.C.E.

PLANET	MASS IN A.U.	SEMI-MAJOR AXIS (MILES)	SEMI-MAJOR AXIS (A.U.)	ENERGY
The Scene - January 1, 701 B.C. E.				
Old Venus	0.814988	67,500,395	0.726156	-22.153944
Old Earth	1.012303	92,339,242	0.993367	-20.115482
Old Mars	0.107447	146,579,410	1.576872	-1.345017
Old Jove	317.799873	483,630,834	5.202804	-1205.718718
			Four Planet Energy Total:	-1249.333160

The Final Fling Between Mars And Venus - Approx. Jan. 24, 701 B.C.E.

Mars gained .086479 energy units

Venus lost .086479 energy units

PLANET	MASS IN A.U.	SEMI-MAJOR AXIS (MILES)	SEMI-MAJOR AXIS (A.U.)	ENERGY
The Scene - March 1, 701 B.C. E.				
New Venus		67,237,928	0.723332	-22.240423
Old Earth		92,339,242	0.993367	-20.115482
Interim I Mars		156,651,477	1.685225	-1.258537
Old Jove		483,630,834	5.202804	-1205.718718
			Four Planet Energy Total:	-1249.333160

The Final Waltz Between Mars And The Earth - March 20-21, 701 B.C.E.

The Earth-Moon System gained .133424 energy units

Mars lost .133424 energy units

(Mars lost twice as much energy as it had gained from Venus).

PLANET	MASS IN A.U.	SEMI-MAJOR	SEMI-MAJOR	ENERGY

An Analysis Of Flyby Distances

In this model of Mars catastrophism, Mars and Venus did a passionate polka on or about January 24, +/- one day. This polka in late January 701 B.C. was, in a gravitational sense, passionate. In an electromagnetic sense, it was also intense.

Ardent as the Ares-Venus dance was, at .086479 energy units, it was only 54.423% as passionate as was the last waltz between Mars and the Earth some 56 days later. The last waltz, even more passionate, was at .133424 energy units.

It had been a long time since the Solar System had experienced anything as dramatic as these events in 701 B.C.E. These polka left Venus closer to the Sun by 262,467 miles. And the last waltz left the Earth-Moon system farther out by 616,565 miles. The combination of these two dances plus modest effects by Jupiter left little Ares closer in to the Sun, by 4,943,437 miles. See Table XIII.

Table XI, further analysis, indicates that the last Mars-Earth waltz, at its closest and most passionate, was at an estimated distance of 27,000 miles, planet center to center. (Since Mars has a radius of 2,100 miles and the Earth 3,950, the two closest surfaces were about 21,000 miles distant).

The last Mars-Venus polka was only 58.423% as passionate as was the last Mars-Earth waltz. Venus is smaller than the Earth; it weighs 81% as much. The Mars-Venus flyby distance planet center to center, is also estimated, at 35,000 miles.

The radius of Venus is 3,800 miles and that of Mars is the aforementioned 2,100 miles. Thus, the nearest distance between the surfaces of these two planets at the closest moment (peri-Venus) was an estimated 29,100 miles (35,000 - 5,900 miles).

Is it any wonder that recent photos from space missions photographing the surface of Venus from nearby reveal a violent, recent physical geography of the Venusian surface? The damage appears as if it happened only yesterday, and to astronomers, a time span of 3,700 years or 12,000 years ago is only "yesterday".

Hesiod and Isaiah both saw the two flybys. Hesiod saw them from Thebes, Greece. Isaiah saw them from Jerusalem. Hesiod wrote that first Ares had a tangle with Pallas Athene (Venus), and next Ares had a cosmic encounter with the Earth. It brought much havoc, lightning, earthquakes, volcanism, etc. He mentions the steeds of Ares, Deimos and Phobos, 17 times in 490 lines.

Hesiod is in disagreement with modern astronomers that Mars, with Deimos and Phobos, has always been 30,000,000 miles or more distant from the Earth over the last 4 billion years. They were close enough to be seen, and as we shall see in chapter 11, their distances were noted in Mars diameters.

Swinging on a dance floor with different partners for different dances often involves in swinging different directions on a dance floor. So it was with these two planetary flings of Mars; they were also in opposite directions (with respect to the Sun.) In the Mars-Venus polka, little Mars was on the inside, being pulled outward by Venus.

Venus, on the outside, was pulled inward, somewhat closer to the Sun. As is mentioned above, Table XIII indicates that Venus ended up shifting some 262,467 miles closer to the Sun, according to calculations of our staff. As a consequence, Venus acquired a permanent increase of 1% in solar radiation. Hot, Venus became even hotter, by 0.36%. Also as a consequence, the Earth, now at 92,956,000 miles, receives 1.49% less solar radiation at 92,339,000 miles than during the Catastrophic Era.

In the process, Mars also attracted and pulled the on-looking, nearby full Moon inward by some 3,070 miles, or 1.28% of its orbit radius. It came in from 30.103 new days (one twelfth of the Earth's old period) to 29.53. One result is that new lunar tides in the Earth's oceans and on its sea shores increased slightly, some 3+%.

Another result was that the ancient 30-day lunar calendar no longer served the ancients at all well. Nor does it serve modern persons well. But the old catastrophic era calendars had to be changed. Every year they were five more days out of synchronization with the Sun, and they kept adding up. For farmers and sailors, the old calendar soon became intolerable. Crops had to be planted correctly, and voyages dated realistically.

One - Ninths --- Three Times

A) Mars, at .107 mass. is about one-ninth the mass of the Earth. The Moon, at a mass of .0123, is 1/81th of the mass of the Earth.

B) Thus the Moon also happens to be about one-ninth of the mass of Mars, one part in 8.7. In the Catastrophic Era, it calculates that the Moon was about 241,900 miles from the Earth. This relationship was invaded, and permanently changed, during the last Mars flyby. However, according to this model, Mars passed through the Earth-Moon system much closer (about eight times closer) to the Earth as to the Moon.

Resonance studies indicate that in the Catastrophic Era, except for the last flyby, all of the 100+ Mars flybys were on the inside, or sunny side of the Earth's orbit. Only the last Mars flyby was unique, on the outside, or night side.

C) It calculates out that, planet center to center, at its closest, Mars was about 27,000 miles from Earth, and simultaneously 214,000 miles from the Moon. Thus, the Mars flyby was at one ninth of the distance between the Earth and the Moon.

The ancients noticed these changed conditions after 701 B.C.E., and soon realized their 30-day lunar calendars were next to useless. Despite evidence, the leaders of 20th century gradualism have rejected the idea of ancient 360-day calendars. This is because they would also have to accept a sudden change in the distance of the Earth to the Sun, and its causing factor, a Mars flyby. For the same reason, those leaders cannot accept a change in the Moon's ancient distance from our planet. It would open the gate to further changes and developments.

The system of gradualism was conceived by various anti-clerical thinkers such as Kant and Laplace in astronomy, and Charles Lyell in geology. With such a predisposition, those philosophers would give no credit to the Bible for its history, much less for its ethics and theology. Now, were they to change, gradualists would need to go back, and read the Bible for its history and geography. They might also gain greater respect for its ethics and theology.

Then they would need to read and assess, Plutarch, Apollodorus, Hesiod, Homer, Heraclitus, etc. No, it is easier to simply reject, and reject again, and be stylishly "politically correct" in an age of compromises of all kinds.

In ancient society after society, there was a clamor by farmers, sailors and mathematicians for corrected calendars. There was a need in Egypt to revise the calendar. The Egyptian astronomical society, met in Canopus in 238 B.C.E., to create a better calendar. Their conclusion was called the Canopus Decree, and it led to a revised 365-day calendar for Egypt.

It was a revision of two earlier Egyptian calendars. One earlier calendar was related to the Earth's spin rate, their 360-day calendar. The other ancient Egyptian calendar was related to the orbit of Venus, in an era when five Earth orbits equaled precisely eight Venus orbits. The Venus calendar of the ancient Egyptians had been the more accurate of the two Egyptian calendars in that era.

It is to be noted that the old Egyptian Venus period is cited at .625 of the Earth's period in Tables XI and XIII. .625 is a perfect 8:5 resonance. The old Venus calendar of ancient Egypt had been a very useful one, and it might be that the ancient Mayas had a Venus calendar somewhat similar to the Egyptian Venus calendar.

For the Hebrews, a simple method was developed to revise the old calendar and still retain its now imperfect twelve month scenario. Merely, it was to add five days to the twelfth month of the year, which was named "Adar". This was done and the additional five-day period was called "Veadar," the new, or "the second" Adar. Veadar made its appearance years after 701 B.C.E. Their Veadar calendar also states much the same thing as Plutarch's assessment of the ancient order, centuries earlier.

However such a simple adjustment as adding five days per year still had an imperfection. With this change, there came to be 36,500 days per century, but the Earth's orbit had 36,525.6365 (new) days. The imperfect Veadar calendar still was off 25+ days per century.

So there was a need for 25 leap days added per century, one every four years. But that still wasn't quite good enough for astronomers. So analysis required for one leap day to be missed once per century (except for every fourth century). The old calendar of the earlier era had major problems, but so did the new 365-day calendar have minor problems, unless perfected.

ELECTRICAL EFFECTS. Another effect of these two swinging events involved the ancient planetary magnetic fields in space. There is no dynamo in the inner core, as most scientists suppose. In a future work of this series, it will be demonstrated that Mars flybys were the dynamo of our planet's geomagnetic field. And it will be demonstrated that the geomagnetic field is housed in the cool iron of the Earth's crust. Our planet no longer has its old dynamo.

As a consequence, its geomagnetic field, like any old once magnetized iron nail, is in the process of magnetic decaying. The Earth's rate of geomagnetic field decay is a half life every 1,350 years.

The Earth's geomagnetic field strength now is .3 Gauss. In 701 B.C.E. as was 1.2 Gauss. In 4700 AD it will be .075 Gauss. Ultimately it will become so weak as to no longer shield the Earth from the Sun's short wave (actinic) radiation.

This is a major long term problem for mankind, and for the future of all flora and fauna. Physicists need to analyze its present decay effects on the biosphere. They need to predict the future effects of the decay. And if possible, they need to find a method of planetary remagnetization. If they don't, the survival of fauna and flora on this planet's surface is doomed within five to seven thousand years.

The ancients in their literature discussed dramatically how destructive Mars flybys were. But Mars flybys were a two sided coin. On the positive note, Mars flybys created strata on continents with sudden oceanic floods; they created suddenly uplifted mountain systems from sudden magmatic crises. In the magma they created mixes of water, silanes and carbides to convert into (l) more silica, or crust, the chemical byproduct, (2) more subsurface petroleum.

Also, paleomagnetic polarity reversals occurred as is evident in basalt formation, and spin axis relocations (shifts) occurred, as is evident in climatology and ancient literature. Isaiah recorded one of them for history, a shift of 10°, and Joshua wrote about another.

Gravitationally, there was a strong mutual attraction (gravity) between two bodies, Mars and the Earth, attracted passionately in space during these swinging events. The gravity of Mars created immense tides in the Earth's oceans and in its sub-crustal magma. Volcanism suddenly erupted from dormant craters on the Earth, and nine times as much volcanism, or more, erupted from the gigantic volcanoes of Mars.

The Earth in close flybys created massive eruptions on the surface of Mars. In this long, eruptive process, one of the Martian volcanoes eventually attained an elevation of 13 miles (Olympus Mons). Its cone covers 110,000 sq. miles and contains 500,000 cubic miles of lava. (If there were 250 flybys, that would average 2,000 cubic miles of erupting lava per fly. And that is merely the largest of the Martian volcanoes. Others rise to 35,000, 40,000 and 50,000 feet. That lava was pumped up by gravities of the Earth and Venus during ancient flybys.

Electrically, it was an attraction, not gravitational combat. In space, the two planetary magnetic fields of old Earth and old Mars BRIEFLY UNITED IN SPACE for ten to fifteen hours. Soon they were torn apart ... as Mars parted company with the Earth ... and both experienced a reversed polarity for the magnetic field of each planet. The south magnetic pole and the north magnetic pole were simply and suddenly reversed. The spin axis sometimes relocated also.

Major league baseball schedules used to feature double headers. The year 701 B.C.E. featured a cosmic double header, two shows in the theater of the cosmos in sixty days. Admission was free --- for the survivors. The first of the two in that remarkable year was "The Last Mars-Venus War." The second feature, even more dramatic (from any perspective) was "The Last Mars-Earth War." Both celestial cinemas were filled with orbit-changing action.

Electrically it resembled something like a brief tryst in space, as Hesiod described. Ares had a sometime "date" with Pallas Athene; this would be the last of them. There was (a) an ion exchange as well as (b) an energy exchange. In addition, there was (c) an angular momentum exchange. As has been mentioned, this one, the last polka of Mars with Venus, was only a prelude to 56 days later, its last waltz ever, the red planet's last dance with the Earth.

Both Hesiod and Isaiah saw the two flyby scenes. As was mentioned, Hesiod viewed the scenes from the Greek Thebes, a place some 40 miles outside the city of Venus. Athens, named after Pallas Athene, a cosmic deity, was the largest city of the province of Attica).

The other viewer-reporter, Isaiah, saw this celestial show, perhaps from a parapet on the city wall of Jerusalem. Each reported and recorded the events they saw for posterity. We, both 20th century catastrophists and gradualists, are that posterity. But they wrote in terms understandable and acceptable to their fellow citizens.

On The Scene Reporting

The opening caption of this chapter cited briefly a section from Hesiod's "The Shield of Herakles." This was a rich reporting job, less than 500 lines of verse about one of the most spectacular scenarios in the history of our planet. The second of the last two wars of Mars was a close brush with the Earth.

As Mars approached the Earth, for a few minutes that night Mars was fully reflective, a "full Mars," and with our Moon also at full moon in the background. When full, our Moon, diameter 2,160 miles, occupies over a half degree (.515) in the night time skies. Mars has a diameter of 4,212 miles, almost double that of the Moon. At 240,000 miles, Mars occupied 1.0057° in the night time sky.

The Moon's reflectivity (albedo) is 7% while the reflectivity of Mars was 16%. Thus at similar distances, and if both were full, Mars was nine times brighter than the Moon (16% x 4 / 7%). When Mars was threatening the Earth at 240,000 miles distant, causing widespread terror, the red planet covered one full degree in the night time skies. It was seven hours (and 213,000 miles) from perigee.

At 120,000 miles Ares covered 2°, both horizontally and vertically and it was only three hours from perigee. The tocsin alarms were sounded, gongs in Japan and India, began to be gonged (54 times, once every few minutes). Trumpets of warning in the cities of the Near East were blown in most cities. Elsewhere other tocsin sirens were rung or were blown, warnings to the populace to find quickly their shelter, in caves, cellars, fox holes, etc. It was three hours to perigee (climax).

At 60,000 miles the diameter of Indra-Horus-Baal-Enlil covered 4° as more populations on this planet became frantic, terrorized, and most had run to their prepared places of refuge. It was a universal epidemic, it was Mars-phobia. It was now one hour and a few minutes to perigee (maximum upheaval). Mega Richter scale earthquakes rattled; dormant volcanoes began to re-erupt.

When at within 30,000 miles (planet center to center) Mars was some 6 minutes from climax. That night, at climax, Mars' diameter covered over 8° of the night time skyscape. Its shining area at its climax appeared to be 256 times greater than the full Moon. Its reflection of sun light was 550 greater than the brightness of a full moon.

This was the "Disc of Mars" (The Shield of Hercules) as it was seen on that night. A round disc was like the shape of Mars. The color of electum was its coloring. Electum was a shiny mixture of silver and gold. When Hesiod titled his work, to him and his fellow Greeks, it meant "The Shining, Electum-colored Disc of Mars". Translators of Hesiod have lacked the perspective of planetary catastrophism, and so they have missed the point in Hesiod's titling of this saga of catastrophism. The "Shield of Hercules" was "The Disc of Mars".

The bright, light orange-colored, surface of Mars reflects sunlight almost 2.3 times more efficiently than does the duller, darker surface of the Moon. In astronomical parlance, surface reflectance of a planet is "albedo." The albedo of the Moon that night was 0.07. The albedo of Mars was 0.16. The silvery, orange planet reflects a full 16% of its received sunlight.

The intensity of tides follows the principle of mass over the distance cubed. At 240,000 miles, Mars' diameter was 1.94 that of the Moon. On the Earth's sea shores, tides generated by Mars were nine times the normal lunar tide. At 120,000 miles, four hours closer, Ares' diameter was 3.9 that of the Moon. Its area was over fifteen times that of the Moon. Its reflectance, if full, was 35 times that of the Moon. This was about three hours before midnight, Athens and Jerusalem time. At that moment, tides on the ocean shores caused by Mars were 68 x the normal lunar tides. And Mars was still coming in.

At 60,000 miles, one hour from maximum, Mars-Enlil's diameter was 7.8 that of the Moon. Its area was 60 times that of the Moon. Its reflectance, if full, was almost 135 times the full Moon. It was barely one hour from midnight.

As was mentioned above, tides in the oceans and in the magma increase according to the equation, mass over distance cubed. Halving the distance produces tides EIGHT TIMES greater in intensity. The following is an indication of tidal intensities as Mars approached:

Oceanic Tides

Mars at:	480,000 miles	1.06 of normal lunar tides
	240,000	8.50
	120,000	68.00
	60,000	544.00
	30,000	4,352.00
	27,000	5,970.00

The Mediterranean Sea is almost a tideless sea since it is entirely enclosed from the Atlantic except for some 8 miles. That is the narrowest part of the Strait of Gibraltar, or, to the Greeks, the Pillars of Hercules. Thus it was that open oceanic shores of the five oceans were treated far, far worse by tides than was the shores of the Mediterranean "Lake".

Mars Diameter

Mars at 480,000 miles	0.97 of lunar diameter
240,000	1.94
120,000	3.88
60,000	7.76
30,000	15.52
27,000	17.24 of a "full moon"

Timing

Mars at 480,000 miles	15 hours to perigee
240,000	7 hours
120,000	3 hours
60,000	1 hour
30,000	6 minutes
27,000	"perigee"

At 30,000 miles, yet closer, Mars-Indra's diameter was 15.5 that of the Moon. Its area was 62.1 of the Moon. Its reflectance, if full, was 140 times the Moon's brightness. This time is estimated as approaching midnight for our two reporters, Isaiah and Hesiot. (Talmudic material indicates the climax was around midnight).

At 27,000 miles, perigee according to this analysis, the diameter of Bel/Baal/Horus was 17.24 times that of the Moon. Its area covered 69 times (2 x 2) as much skyscape compared to the Moon. Its reflectance was 157 times the brightness of the Moon if both were full. This was the perigee of Mars that night, perhaps at a few minutes after midnight. Oceanic tides and subcrustal magmatic tides were 5,900 times normal Moon-created tides, hence the crescendo of massive earthquake and volcanic activity.

Some were wondering if this would be the end of the world. Or of the Earth. No. But had they understood Roche's Limit, Mars had come within 16,000 miles of shattering and being turned into countless asteroid fragments. (Roche's Limit for fragmentation is 2.44 radii and the Earth's radius is 3,900 miles). Had Mars come within 11,000 or 12,000 miles of the Earth, it would have fragmented just like Astra, many millenniums earlier.

Considered together, Ares, "bane of mortals" (Homer), made a close flyby, as it rapidly closed in on the Earth. (To Greeks, "It" was a "He", Ares, the celestial warrior. Athena, or Venus was feminine). Mars was accompanied by its celestial steeds, Deimos and Phobos, on full display, just as Hesiod and others reported.

Hesiod referred to one or both of the celestial steeds of Mars, Deimos and/or Phobos, in one form or another 17 times in only 486 lines. These include lines 97, 144, 154, (Onrush and Backrush) (Battle noise and Panic), 195 (Terror and Panic), 347, 370, 372 (fluttering-maned horses) and 463 (Panic and Terror).

Tiny Deimos has fragment diameters of only 6 x 7.5 x 10 miles. Phobos was a little larger fragment, with a maximum diameter 17 miles. Hesiod viewed the orbiting of these tiny asteroids as the turning of the wheels of the chariot of Mars. Phobos has a period of 7 hours 29 minutes. Deimos, 30 hours 21 minutes. (Gradualists note, Hesiod could not have seen them if Mars were 2,000,000 miles distant).

For future generations, posterity, (including us), Isaiah and Hesiod reported the events of that night. It was a scary, close, midnight Passover of Mars, overhead, to a surrounded, very frightened city of Jerusalem, a city that happened to be crammed with refugees.

Assyrian assault armies were waiting, a mile or so beyond the city walls. Sennacherib and his star gazer advisors viewed this coming scene as an opportunity rather than as a catastrophe. Lightning might hit Jerusalem, fry it and blow it down from the inside. Or, more likely, earthquakes might compromise the integrity of its walls, and thereby simplify the coming brutal assault.

From such a sensational starry scene, including a visible Deimos and Phobos, a person might succumb to either dementia tremens, cometophobia (about which more is said in chapter 1), or "Ares-phobia." (Deimos and Phobos were two words among some two dozen Greek words for different kinds or categories of fear).

These events were both foreseen and reported by Isaiah in the Book of Isaiah in the Bible, chapters 1 to 38. Also they were reported in Hesiod's rich 486-line account which he titled "The Shield of Herakles (Hercules)." They are also reported in Ginzberg's "The Legends of the Jews".

Incidentally, the title of Hesiod's work seems to have suffered from mistranslations by later translators, scholars who never imagined planetary flybys. As was mentioned above, a more accurate title, which Hesiod probably had in mind, was "The Disc of Ares". A shield was a disc and Hercules was among the archetype names of Ares, as was Phaethon, Gorgon, Lotan, Medusa, Perseus, etc.

Hesiod

In Greek mythology, Hercules was very strong; he was one of many archetypes of Mars in Greek literature. Ares may have gotten a new nickname with each flyby occasion. Other archetypes included the swift Perseus, the evil Medusa, the hideous Cyclops, the even uglier Gorgon, the ravaging Lotan, etc. Essays by David Talbott and Ev Cochrane consider that Ares had 110 archetypes or "nicknames" in Greek literature. [n8], [n9]

Lotan, one of the Greek archetype of Mars, is derived from the Chaldean language. This Chaldean word was borrowed or inherited by the ancient Hebrews and appears in the Bible as Leviathan, the celestial dragon in the Book of Job, ch. 41.

With some 100+ archetype words, all descriptive words for Mars in Greek, this is a measure of the swiftness, the luster, the terror and the repeated damage wrought by a numerous close Mars flybys in ancient times. In another volume evidence will be cited that the Mars flybys were in cycles, 108-year periodic cycles, periodic to the day and virtually to the hour.

The great monsters or dragons of the celestial deep were known by the Hebrews, as by the Chaldeans, under several names. Those names included Leviathan, Behemoth, Teammate, Asp, and Rehab. Their assaults on the Earth go back to the era when these celestial assaults began, when the Earth's geomagnetic field was created, and when paleomagnetic polarity reversals began to appear in lava flow sequences. Leviathan, in our analysis of Chaldean thought, appeared in October flybys, while Behemoth was seen and feared in Chaldean March case flybys, or "Passovers".

The scenery, that frightening night, stirred both Isaiah and Hesiod to record it. Since modern translators have no idea of planetary catastrophism, they have chosen a lesser, somewhat softer terms, non-astronomical terms for celestial scenes they could not understand. As was mentioned earlier, round, electum-colored discs became shields. They can be reflective if made of iron or bronze and polished. Electum was a shiny alloy of silver and gold. The shield of Mars according to Hesiod had the coloring of an electum disc.

Shields were used in earthly combat. Circular shields were involved in celestial combat. This night a round, shield-like disc was involved in intense celestial combat with the Earth. It was with the disc of Ares.

> For all about the circle of it, with enamel and with pale ivory, and with electrum it shone, and with gold glowing it was bright, and there were folds of cobalt driven upon it.
>
> In the middle was a face of Panic, not to be spoken of, glaring on the beholder with eyes full of fire glinting. and the mouth of it was full of teeth, terrible, repugnant, and glittering white, while over the lowering forehead hovered a figure of Hate, marshaling the slaughter of fighting men, cruel spirit, who took the senses and perception out of those fighters who tried to fight in the face of Zeus' son, the War God. [n10]

As the catastrophe deepened, there was a thick smoke and ash smog in the air from burning forest fires and volcanic eruptions. The next morning the smoke and ash smog in the atmosphere altered the color of this light silvery orange disc into a deep red. Forest fire fighters readily understand this coloring due to a thick haze of smoke in the atmosphere, changing yellow sunlight into a red glow.

It was a night to be remembered, so Hesiod wrote in order that it could be remembered by future generations. He had no idea this flyby would be the last of a long series of waltzes of Ares with Hera. So it is that modern translators of Hesiod, unfamiliar with planetary catastrophism and the unruly Ares, thus have had to struggle in translating the scenes Hesiod described.

Interestingly, in his report, Hesiod portrayed Mars as first tangling with "gray-eyed" Pallas Athene - Venus. Apparently, interplanetary discharges were generated by fierce friction in the subcrustal tides of magma. Between Ares and Pallas Athene, the bolts of lightning lit up the cosmos, including lighting up the cometary tail of Ares. This was some 56 days before the Final Flyby.

The reflective phenomenon, of discharges on Venus lighting up the cometary tail of Ares, was a rare phenomenon for the ancient Greeks. Mars-Venus flybys occurred rarely, but when they did, cosmic lightning lit up the cometary tail of Mars like a Christmas tree. It was an "aegis". Modern translators of Hesiod into English have struggled to find a word for Ares' cometary tail.

It was a reflective, silvery, effervescing celestial gauze, reflecting cosmic lightning, interplanetary Venus-Mars discharges. In English there is no word other than the Greek "aegis" for this phenomenon when lit up. Otherwise the icy, cometary tail of Mars it was known as the Fleece of Aries."

> It was a wonder to look at, even for Zeus deep-thundering, through whose counsels Hephaistos had made the shield, great and massive, fitting it with his hands.
> And now the powerful son of Zeus swing it with full control, and leaped down from the horse-chariot like a lightning-flash from the hand of his father,
> Zeus of the aegis, stepping light on his feet, and his charioteer, strong Iolaos, standing firm on its floor steered the curved chariot.
> Meanwhile, the goddess, Athene of the gray eyes, came and stood close beside them ... [n11]

So modern translators settled upon the ancient Greek term, the "aegis" of Ares. Only, no translator today can define "aegis" except that it may have had something to do with Pallas Athene and the warlike son of Zeus.

In addition, Hesiod, in his brief 480 lines in "The Shield of Herakles," did not overlook the appearance of Deimos and Phobos as the two black celestial steeds of Ares. Adjective and noun phrases used by Hesiod, and now translated, include such terms as "Shaker of the Earth" (line 104), "the War God" (150), "Manslaughter" (155), Grim-faced Ares (191), Death mist (264), "Man slaughtering Ares" (334), "Lord of Battles" (371), "Son of Zeus", (391), "Stout-hearted son of Zeus", (424), "Man slaughtering Ares" (425).

These black steeds were described drawing the shining, warlike, sword-swinging chariot of Ares across the celestial scene, the cosmos.

As was mentioned, Deimos and Phobos, and words for them occur some 17 times or more in 486 lines. Panic and Terror. Swift-footed horses. Fast horses with reins slackened. The horses that drew his chariot. Horses on either side. Fast-footed horses. Fluttering-maned horses. Two of them.

Isaiah and Sennacherib

Isaiah, like Hesiod, was a reporter of the last cosmic waltz. Among his predictions were forecasts of [high voltage] crashing cosmic lightning discharges, gigantic killer shock waves, terrible claps of thunder [heard for thousands of miles], rocking earthquakes, vigorous volcanic eruptions and an unsettling of the cardinal directions, hence a spin axis relocation. All of these had occurred during the flyby of 756 B.C.E., perhaps when Isaiah was a child observer.

The earth is utterly broken down, the earth is clean dissolved, the earth is moved exceedingly. The earth shall reel to and fro like a drunkard and shall be removed like a [summer] cottage. Isaiah 24: 19-20

Drunkards reeled; so could the spin axis of the Earth. Summer cottages could be easily relocated with a pair of staves and a sextet of pole carriers. So also, the North Pole, the South Pole and the equator were relocated. Isaiah had it right. It was somewhat of a repeat of the 756 B.C.E. catastrophe, except it was on a Passover anniversary in the spring. It was at night rather than during the daytime. But his language was good for the Hebrews of his era, though it was not well tailored for modern science.

Isaiah first predicted, and then reported on the next day after the Final Flyby, as if "the spin axis had shifted." Isaiah reported it in a factual manner for the citizens of Jerusalem. He recorded the shadow on the Jerusalem Sun Dial shortened "10°." Someone had measured the new shadow versus the previous shadow, probably as cast at noon, at the Sun's zenith.

The standard of measure he used was lost by the Jews after their Babylonian Captivity, when they lost their Hebrew language. So, translators, in Hebrew, Latin and English, have guessed and selected the English word "degrees."

Probably the original term related to a measurement of angles. The new versus the old angle was measured and the shadow was shorter. Apparent the shadow was ten "short cubits" shorter, our educated guess. And the obelisk was perhaps 75 feet high, again a guess. Angles and degrees are related and this job of translation is fairly good.

Sennacherib's astrologers, close to the throne, had advised him that the Assyrian army ought be stationed outside Jerusalem at least by our date of March 20. On that night, earthquakes would test the defensive walls, and perhaps shatter them, or at least damage them. And one bolt of cosmic lightning from the Assyrian Nergal (the Chaldean Bel, the Greek Ares) might turn the holy city into a flaming holocaust. Thus, Sennacherib's astrologers recommended taking Jerusalem the easy way, enlisting assistance from Nergal-on-high.

With this vast army Sennacherib hastened onward, in accordance with the disclosures of the astrologers, who warned him that he would fail in his object of capturing Jerusalem if he arrive there later than the day set by them. [n12]

The Jewish Talmud also records the scenery and the dramatic history of that unusual and pivotal night. Sennacherib wanted to capture Jerusalem, either by surrender or by sack. If by sack, it meant the total slaughter of Jerusalem's inhabitants, a holocaust, Nazi-like, including all stray pets. If Jerusalem was taken by surrender, it meant deportation of the surviving populace to a cold, remote northern land, the land of the Volga, Southern Russia, 1,000 miles, or elsewhere in cities of remote Inner Asia.

King Hezekiah anguished over Sennacherib's terms of surrender, and asked for Isaiah's advice. Isaiah advised faith and resistance; the Lord would deliver Jerusalem. Taking Isaiah's advice, Hezekiah resisted, a chancy decision indeed in light of the record of Assyrian annihilation of cities elsewhere in the Near East.

The "Senna" part of Sennacherib's name means "the Moon." The "cherib" part of his name in Semitic languages is another word, as shall be demonstrated, for "the cosmic marauder, or Mars." The translation in English is "cherub," which was a destructive angel or messenger of the Lord. This is discussed in more detail in Chapter 12. Those same two celestial bodies (Mars and the Moon) "by chance" were the two bodies whose orbits were reduced in diameter that very night.

That night, March 20, 701 B.C.E., Sennacherib's well-armed panzer, with their ample iron armor, were on schedule, and on duty, camped outside the western wall of the city, some 250,000 strong. Sennacherib's terms, as usual, were "either-or". It was either (1) to surrender the city and accept deportation of the entire populace into a far away land, abandoning forever the land of Israel, or (2) to suffer total annihilation. No more children of Israel. It was reminiscent of Hitler's style.

In the Bible, the events of that night are found in the Book of Isaiah, chapter 38, and in two other places of the Old Testament, in II Kings and in II Chronicles. Further description at greater length has been recorded in the Talmud, cited as follows.

The archangel Gabriel, sent by God to ripen the fruits of the field, was charged to address himself to the task of making away with the Assyrians, ...

The death of the Assyrians happened when the angel permitted them to hear the "song of the celestials." Their souls were burnt, though their garments remained intact. [n13]

Hezekiah and Isaiah were in the Temple when the host of the Assyrians approached Jerusalem; a fire arose from amidst them, which burned Sennacherib and consumed his host.

See Tehellim 22, 180. The burning of Sennacherib is not to be taken literally. [n14]

"That day" was the night of March 20-21, when a close flyby of Mars would create subcrustal tides, crustal earth shocks, volcanic upheaval and interplanetary mega-thousand volt electrical discharges on the Earth's surface. Sennacherib accepted the advice of his astrologers and was there on schedule, with his abundant arsenal of armor, much of it made of Hittite iron.

Iron is superior in hardness to other metals, important for military purposes. But, as Benjamin Franklin discovered, iron also attracts lightning, and therefore it is an excellent material for manufacturing lightning rods. An arsenal of iron armor makes an even better attraction than does a handful of iron weapons. Iron's magnetic properties were not realized by either the Assyrians or Hebrews, the Greeks, the Trojans or the Romans. It is understood only by modern scientists and their students.

The ancient rabbinical commentators wrote for the benefit of later scholars, and to our modern age of gradualism. Later Hebrews lost the ancient literal perspective of catastrophes, having never seen one, and having lost the original Hebrew language. Never would they experience interplanetary electrical discharges, on scales of 220 volts, much less on scales of 10,000 to 50,000 volts. Therefore they tendered that the burning of Sennacherib on one side of his face "was not to be taken literally." They were mistaken. This blast in some ways was like a small atomic bomb.

There were some eight or nine kinds of electrical effects and phenomena that occurred during March flybys. The Earth's geomagnetic field swept across the cosmos and briefly, for 10 to 20 hours, united in space with the planetary magnetic field of Ares, or Mars. [n15]

The ancient, but post Babylonian rabbinical commentators were trying admirably, but came not even close to an understanding. Thus they rationalized that the lightning blast which originated from the surfaces of Mars and the Earth instead originated in the Earth's troposphere, or its stratosphere. Actually their ancient, original sources had been quite correct, and they ought to have been taken literally, without an added disclaimer.

> Sanhedrin 95b, and similarly Jerome on Is. 10.3. The latter states that Jewish tradition considers Hamon, "noise" (comp. Is. 33.3), to be the name of the angel Gabriel. This is corroborated by Aggadat Shir 5, 39. According to Sanhedrin, the angel clapped together his wings, and the noise caused by it was so terrific that the Assyrians gave up their ghosts. Another view given in Sanhedrin is that the angel blew out the breath of the Assyrians. This means that he took their souls without injuring their bodies. [n16]

In response to the high voltage lightning strike, there was a killer shock wave. Apparently there was a flash somewhat comparable to a small atomic bomb. Sennacherib himself was reported as being a survivor, but with burns - flash burns, perhaps on one side, like Hiroshima.

Immediately following the blast of lightning came the killer shock wave. Traveling in the Earth's atmosphere, it was heard for thousands of miles, perhaps as far as the Western Hemisphere. The celestial lightning had struck, but not in the citadel of Jerusalem, as it might have; it struck Assyrian heavy iron armor.

In three separate citations in the Bible, (Isaiah 38, II Kings and II Chronicles), the mortality count was recorded by that one monumental, massive, memorable nocturnal blast just outside Jerusalem - 185,000 dead Assyrian troops. What was left of the scared and scorched Assyrian army beat a hasty retreat to Nineveh. There, Sennacherib, burned and a big loser, was assassinated.

Thus it was that the "angel Gabriel," a destructive and dreaded messenger of the Lord indeed, delivered Jerusalem. It had happened once again. Something similar had occurred on the Long Day of Joshua, some 701+ [or 702] years earlier. [October, 1404 B.C.E.].

Something similar had happened in the years 1296, 1188, 1080, 972, 864, 756 B.C.E. and a half cycle later, in March of 701 B.C.E.]. Catastrophes were in 108-year cycles, as if they had something to do with Jupiter as well as with Mars. (The Earth was in 12:1 resonance with Jupiter, and l08 is divisible by 9). Thus Jupiter was always at the same place in the cosmos when catastrophes hit.

The above citations and others uncited ones in the Talmud indicate there was a record of the "deliverance's" although not well understood. Sennacherib's song, a rabbinical theological version of the deliverance of Jerusalem, and the final flyby, are all one, describing the last waltz of Mars and the Earth.

The scientific version of 20th century planetary catastrophism is that first of all, Sennacherib had no business invading Judah or Egypt. He had no business uprooting and deporting entire populations, ethnic cleansing, which had occurred to Northern Israel and other populations elsewhere the Assyrian Empire. But that is by our contemporary standards, not his.

Further, if he went ahead anyhow, Sennacherib had no business ever centralizing his iron armor in his encampment. Especially he should not centralize his iron armor anywhere in a year and in a month and week when the Assyrian "angel of the Lord," Nergal by name, was to make a close Nergal flyby.

Clearly the approaching "angel" or "messenger of the Lord" was visible and it was on schedule according to both Assyrian astrologers and Isaiah. It was on the anniversary of the Hebrew Passover, and earlier catastrophes, such as the Exodus event, also the Sodom-Gomorrah event.

Sennacherib's basic problem appears to have been lust and egotism; he wanted to destroy the armies of Egypt and to plunder the great riches of the land of the Nile. Jerusalem was merely a burr in his saddle on the way, but what a burr.

In our era, ongoing tidal friction within Jupiter creates discharges from Jupiter all the way to the surface of nearby Io, 260,000 miles away. A constant flow of electricity occurs in voltages up to 400,000 - and some 5,000,000 amps - 2,000,000,000,000 (two trillion) watts. This voltage between Io and Jupiter equals 70 X all of the human electrical generating capacity on the Earth.

By comparison that night, Mars was distributing (by Io standards) electrical discharges across only some 27,000 to 50,000 miles of space. The discharges were scattered broadly across the face of the Eastern Hemisphere that night. Perhaps those discharges were a mere 50,000 or 75,000 volts, or perhaps they were of a higher voltage. That Troy had been hit sometimes is established by the way its mortar is fused in its city walls. No other explanation suffices. Until more research is done, their various voltages are unknown.

Sennacherib's army was a casualty. It was much like the Greek army, 162+ years earlier, which had suffered heavy casualties from celestial fire just outside the walls of Troy on schedule, on October 24, 864 B.C.E.

Perhaps this particular bolt in 701 B.C.E. was destined generally to discharge somewhere in the land of Israel, even possibly near Jerusalem. But (thanks be to Sennacherib) it honed in on the fine Assyrian concentration of Hittite iron armor instead, an arsenal of lightning rods.

THE AFTERMATH. In the aftermath of the discharge, the first job was to bury the dead, a big job. Next, the good news, very good news indeed, was sent out far and wide. Scientific types of the late 8th century B.C.E. and early 7th century B.C.E. came from far and wide to inspect the meltdown of iron armor and the crater left at the point of the discharge. There was a small raised mound left at the very center of where the lightning discharged. (So it is with craters on Venus today).

Isaiah became famous as a great prophet, one of the greatest Israel ever had. Hezekiah acquired the reputation as a great and wise king, which he was.

Among the visiting scientific groups was one from Babylon, from some 700 miles to the east. The chairman was one Merodach-baladan, the son of the king of Babylon, Baladan. Baladan, the king was named after Bel, the Chaldean Ares.

Merodach was named after the Babylonian deity Marduk, Jupiter. Thus the name of the prince, Merodach Baladan means "Jupiter-Mars," or perhaps "Jupiter father of Mars." King Belshazzar and the studious Belteshazzar had been named after Mars; Nebuchadnezzar was named after fast-moving Mercury. In Chaldea, naming royalty after the planets was in vogue in the Catastrophic Era.

In this age of gradualism in the cosmos, nobody names their son Mercury, Mars or Jupiter. Only rarely in this culture is a baby daughter named "Venus." But when this does happen, the position of Venus in the cosmos is hardly what the parents have in mind.

In the centuries after this event, some like Heroditus explained the Assyrian debacle as a sudden plague of flea-infested mice or rats, bringing biological disaster to the Assyrian camp in a single night. It was a sincere try, but not a very good one.

Many theologians stick with Heroditus and his mice and fleas with bubonic plague for their explanations. None of them have had the proper base of knowledge to comment.

A few months, even years earlier before the Final Flyby, there were recorded vignettes from what the prophet Isaiah had foreseen and forecast:

> Thou shalt be visited of the Lord of hosts with thunder, and with earthquake, and great noise, with storm and tempest, and the flame of devouring fire.- Isaiah 29:6
>
> I will sweep with the besom of destruction, saith the Lord of hosts. - Isaiah 14:23

It is well known that Isaiah credited the Lord, God of Israel, with their collective good fortune. Apparently the Lord had provided Isaiah with some kind of precognition of what would happen, and why Hezekiah should refuse to surrender. Isaiah told his king that danger indeed was coming but it was danger to the Assyrians, not a danger to the Hebrews. Isaiah's precognition was shades of Noah's, precognition, almost 1800 years earlier.

Conclusion

A logical pattern of energy shifts is offered for the energy conversion of Mars from its Catastrophic Era orbit to its modern orbit.

Story 26 is that the year 701 B.C.E. FEATURED TWO FLYBYS, FIRST, MARS-VENUS, AND SHORTLY THEREAFTER, MARS-EARTH. The first flyby of Venus by Mars was on January 22, 701 B.C.E. The second flyby was on the night of March 21, also 701 B.C. This was only 56 to 58 days later. The energy Mars gained from the first flyby is why Mars went into an orbit farther out, and it went on the wrong side of the Earth during the second flyby. The first flyby was approximately 48% as energetic as the second. This first time, Mars gained energy; it was an exchange. Venus lost an equal amount of energy.

Story 27 is that of the second flyby, THE FINAL WALTZ, WHICH WAS THE MORE

ENERGETIC OF THE TWO FLYBYS. Mars passed between the Earth and the full Moon, on the Earth's outside, or dark side. This occasion increased the Earth's orbital distance from the Sun by 616,565 miles. Our planet's spin rate increased 0.452% (360 to 361.628). Its period increased 1.003% (361.628 to 365.256). The combination of the two resulted in an increase of 1.46% in day count per year (360 to 365.256).

After this siege of catastrophism, the Earth no longer was in 12:1 resonance with Jupiter's orbit, nor was Mars in 6:1 resonance. The orbit of Venus shrunk from .625 to .6152 in period. Venus left an orbit that had been in 8:5 resonance with the Earth. As is shown in Table XIII, chapter 10, the period of Venus diminished by 1.32 day.

Story 28 is that THE MOON ALSO DEPARTED FROM ITS EARLIER 12:1 RESONANCE AND 30-DAY PERIOD. Its new orbit became 29.53 new days rather than the old count of 30.

This chapter addresses the shortening of the SEMI-MAJOR axis of the three bodies, Venus, Mars and the Moon. In three stages, the "x" axis of Mars was shortened from 146,579,410 miles to 141,635,973 miles. This is the new semi-major axis, the "x" axis, the longest axis of the Martian orbit. The new period of Mars became 687 days, down from an old 720 (old count).

The "y" axis of an orbit, the SEMI-MINOR axis, measures the width of an orbit. Changes in the "y" axis measure changes in angular momentum. Chapter 9 addresses the changes, in three steps, of the "y" axis of Mars. It has been deduced that the orbit of Mars rounded out; Table XIII will explain why.

Story 29 demonstrates HOW AND WHEN THE EARTH'S ORBIT ENLARGED FROM A 360-DAY OLD ORBIT CONDITION TO A 365.256-DAY NEW ORBIT CONDITION. Simultaneously, the Moon's lunation, or synodical period shrunk from a 30-day period to 29.53 days. Tables XI, in this chapter, and XII and XIII, in chapter 11, indicate that these shifts were in accord with celestial mechanics.

Such information would have brought a smile to the face of Johannes Kepler (1571-1630), the father of celestial mechanics. No doubt, Isaac Newton among others in London's John Bull club, also would have smiled. So would Jonathan Swift, Edmund Halley, John Arbuthnot, William Whiston and others. Chapter 10 indicates why Arbuthnot, Swift and others would have smiled if they had this information.

Story 30 demonstrates HOW THE PERIOD OF THE ORBIT OF MARS SHRUNK FROM THE FORMER 723.257-DAY CONDITION (OLD DAYS) TO THE MODERN 686.979-DAY CONDITION. This is a major reduction in period of 5.016%. This reduction in the energy of Mars was attributable primarily to the Earth. The three planets causing the new orbit of Mars and their energy exchange were:

Venus (- 184.22%)
Earth (+284.22%)
Jupiter (+ 0%)

This is an indication as to how intense, or astronomically passionate was the final Mars-Earth waltz.

By comparison, the Earth's orbit enlarged from 361.628 new days (or 360 old days) to 365.256 new days, an increase of only 1.003%. Mars, in stage # 2, lost just as much energy as the Earth gained. The Earth's orbit expanded 616,565 miles in its "x" axis. In the Martian "x" axis, there was a reduction by 15,015,507 miles. This was because tiny Mars has only one-ninth of the Earth's mass.

The orbit of Venus shrunk from 226.018 days to 224.700 days, a shrinkage in period of 0.58%. The Mars-Venus flyby featured Mars on the inside. Venus is smaller than the Earth, and the flyby of Mars wasn't quite as close. The Mars-Venus polka was only 64.82% as energetic, as passionate astronomically, as was the last Mars-Earth waltz.

Half of this story of exchanges is now completed, the energy exchanges. The other half, the angular momentum shifts, is in chapter 10. It is an analysis of the changes of the "y" axes of these four planets. The "y" axis, the semi-minor axis, measures an orbit's maximum width.

If the totals of both energy shifts and angular momentum exchanges agree, and agree simultaneously at all stages, this theory will constitute a viable theory of catastrophic cosmology and of Earth history. It will indicate that the popular menu for Solar System genesis and history, gradualism, is the wrong menu.

Unfortunately, at stake is the turf (paradigm) which has been taught as truth to 99.9+% of the 19th and 20th century astronomers when undergraduates. Without realizing it, they themselves became victims of these doctrines and dogmas that were developed (in an anti-clerical atmosphere) two centuries ago.

Each day, unwittingly, these persons, once undergraduates, now perpetuate the foolishness of gradualism, and create more victims. The teacher, at least in science, is always thought to be right, that is, except for the mistaken teachers of the Ptolemaic system 400 years ago. Some of them wanted both Copernicus and Kepler to be burned at the stake, and Galileo as well.

So most embrace these dubious dogmas of the anti-clericalists of 200 years ago. For some, questioning these 200-year old "givens" is painful. So it is that both teachers and students affirm that Mars has always been only a tiny pinpoint of light in the night time sky for the past 4+ billion years. The red planet never ever has been closer to the Earth-Moon system than the modern 33,000,000 miles.

These scholars however pay no attention to ancient history, to ancient calendars, to ancient literature or to the repeated accounts of planetary catastrophism in the Old Testament. In so doing, they have no concept of how much they have missed.

Story 31 is that both the Book of Isaiah and Hesiod's "The Shield of Herakles" WERE WRITTEN IN THE AFTERMATH OF BOTH THE FINAL FLING OF MARS WITH VENUS, AND THE FINAL WALTZ OF MARS WITH THE EARTH.

Add to this Hesiod's "Theogony" and the writings of Apollodorus, plus those of the "fire and brimstone prophets" Amos, Joel, Micah, etc. Other materials include Homer, Ovid, Plutarch and others not discussed here in detail. Vignettes only from Hesiod and Isaiah have been incorporated to widen the perspective of general readers, scientists, historians, geologists, archaeologists and theologians, among others.

Story 32 is THE NATURE OF THE DEMISE OF SENNACHERIB'S ARMY, AND OF THE SURVIVAL OF JERUSALEM. Jerusalem survived the onslaught of the Assyrians, spearheaded by Sennacherib's panzer of 701 B.C.E. Loosely speaking, the crisis night was a near carbon copy of goings on during the Long Day of Joshua, 700 years earlier. At that time, also via catastrophism, the Hebrews had been spared and had triumphed over the larger, well-armed armies of a Canaanite alliance.

The 20th century, and soon the 21st century, is fortunate to have the records of these two such eye witness reporters of that celestial scene. One was from just outside Athens, and the other from inside Jerusalem. The reports of both deserve careful analysis. As was mentioned above, other writers of the era, Greek, Hebrew and beyond, who described catastrophism, also merit attention.

Story 33 is that THERE ALSO WAS A MODEST, INWARD SHIFT IN THE ORBIT OF VENUS DURING LATE JANUARY OF THE CATASTROPHIC YEAR 701 B.C.E. The inward shift of Venus plus the outward shift of the Earth spelled the doom of the ancient Egyptian Venus calendar. The slight increase in the Earth's spin rate coupled with an enlarged orbit insured the doom of the 360-day calendars.

With story 33, the reader now is a full 70% of the way to understanding the catastrophic scenery of ancient times. The view of both history and the cosmos keeps getting better and better, richer and more spectacular.

With respects to Plutarch, Hesiod, Homer, Apollodorus, Isaiah, Joel, Amos, Hosea, Jonah and others, in modern cosmology the score now is: Ancient literary catastrophism - 4, Modern 20th century gradualism - 0.

CHAPTER 10

The Angular Momentum Exchange

The battlefield was darkened by Ares. ...Hera, the goddess of the earth, "stepped upon the flaming car" and self-bidden, groaned upon their hinges the gates of heaven which the Hours had in their keeping, to whom are entrusted great heaven and Olympus. She spoke to Zeus:

"Zeus, hast thou no indignation with Ares for these violent deeds, that he hat destroyed so great and so goodly a host of the Achaeans recklessly? ... Wilt thou in any wise be wroth with me if I smite Ares?"

And Zeus replied: "Nay, come now, rouse against him Athene ... who has ever been wont above others to bring sore pain upon him." So came the hour of battle.

Then Pallas Athene grasped lash and the reins, and against Ares first the speedily drave ... Athene put on the cap of Hades, to the end that the mighty Ares should not see her.

Ares, BANE OF MORTALS, was attacked by Pallas Athene, who sped the spear "mightily against the nethermost belly."

The brazen Ares bellowed loud as nine thousand warriors or ten thousand cry in battle, when they join in the strife of the War-god.

Even as a black darkness appeareth from the clouds when after heat a blustering wind ariseth, even in such wise ... did brazen Ares appear, as he roared amid the clouds unto broad heaven.

In heaven he appealed to Zeus with bitter words of complaint against Athene: "With thee are we all at strife, for thou are father to the baneful maid, whose mind is ever set on deeds of lawlessness. For all the other gods that are in Olympus are obedient unto thee ... but to her thou payest no heed ... for that this pestilent maiden is thine own child."

And Zeus answered: "Most hateful to me art thou of all gods that hold Olympus, for ever is strife dear to thee and wars and fighting.

Homer [The Iliad]

The Background For The Trojan War

Homer, reported to be blind, nevertheless by conversing with his colleagues, he penned a profound and spectacular account of Ares-Hera catastrophism. It was the year 864 B.C.E., a catastrophic year. It was early autumn.

The Greek generals planned an assault on Troy, hoping that either Troy would be struck by celestial lightning, torching the citadel, and/or its walls of defense would be crumbled by killer quakes. An assist by the destructive Ares had been predicted by the Greek star gazers and monthly prognosticators to the Greek military leaders. They seem to have understood the 108-year cyclicism of Mars flybys, and the last October flyby was in October, 972 B.C.E. An easy, quick conquest was anticipated. (It wasn't either quick or easy as it turned out).

Interplanetary discharges of mega-volt lightning struck the Greek encampment and its iron armor instead. The walls of Troy held firm. This campaign for the opportunistic Greeks, bent on maritime expansion into the Euxine Sea region and the Ukraine, suddenly turned difficult, long and a drawn out endeavor.

In the mega-catastrophic year 701 B.C.E., the Greek reporter on the scene was Hesiod and the Hebrew prophet-reporter was Isaiah. In the mega-catastrophic year 864 B.C.E., the Greek reporter was Homer, a blind writer. The Hebrew prophets on the scene were Elijah and Obadiah. Cosmic lightning hit at least one place in Northern Israel that day too. It was a discharge that struck Mt. Carmel, where there were quarters of beef offered in a sacrifice; it resulted in a well-done barbecue. See I Kings 18:38.

As was seen in Chapter 9, iron was excellent material for weaponry in normal years, but the Greeks had not yet learned that the employment of concentrations of iron armor was a hazardous deployment in catastrophic years. Homer described the residual fires in the Greek encampment, burning long into the night after that catastrophic daytime conflagration on October 24, 864 B.C.E.

With its concentration of iron armor, the Greek camp was decimated by high voltage fire falling from the heavens, the same treatment they hoped would smite Troy, as it had done on past flyby occasions. Such was the timing of the opening stage of the Trojan War. Its timetable no doubt had been scheduled by the seers, the astrologers and the monthly prognosticators of Greece, who held seats close to kings and generals. Advice from the monthly prognosticators aside, it was a Greek disaster, directly attributable to Ares, "the bane of mortals" (Homer).

On March 20, 701 B.C.E., some 162+ years later, as was described in chapter 9, the Assyrian army, some 250,000 strong, prepared to attacked Jerusalem, expecting assistance from Nergal to visit killer quakes and/or celestial lightning on the Jewish citadel. Once again the Assyrian arsenal of iron armor functioned as an attracting lightning rod. The iron-abundant Assyrian encampment, like at the Greek encampment 162 years earlier, was struck instead by "fire falling from heaven". The nearby citadel was spared.

Introduction

The Mars flyby of the Earth, occurring in October, 864 B.C.E. was followed by another October case Mars flyby slightly over a century later. Close Mars flybys occurred in 108 year cycles (to the day). The next October flyby would be in 756 B.C.E. The foundation for these dates will be laid in a later Volume, and occur in our essay entitled "The 108-Year Cyclicism of ancient Catastrophes". [n1]

Mankind is fortunate to have this flyby of 756 B.C.E. described in Hebrew style by three prophets of fire and brimstone reputation. Each was a reporting, or recording prophet. They are Jonah of whale fame, Joel of fire and brimstone fame, and Amos of earthquake prediction fame. The flyby of 756 B.C.E. caused great structural damage to the First Temple in Jerusalem, which is cited by Josephus. [n2]

This event in 756 B.C.E. is the beginning point of the three-stage shift of angular momentum in the next catastrophic scene, that of March in 701 B.C.E. The 756 B.C.E. flyby was the last of the autumn (October) waltzes between Mars and the Earth. Peace was to come to our planet and to Venus 54+ years later.

The energies exchanged, described earlier, involve measurements of the "semi-major axes" of the three orbits of Venus, the Earth and Mars. As was mentioned, the SEMI-MAJOR axis of an orbit is half of the length of the long axis of the orbit. It is measured from the center of the orbit to either the aphelion or the perihelion. This line, the "major axis," is also called the "x" axis.

In contrast to the "x" axis, there is a "y" axis AT 90° to the "x" axis. Half of it, from the orbit center to one edge, is the SEMI-MINOR axis. It is drawn at the center of an orbit, and perpendicular to the "x" axis. It bisects the "x" axis. The Minor Axis measures the maximum width of an orbit.

Both the "x" axis and the "y" axis intersect at the center of the orbit. Only the "x" axis is relevant to measuring a planet's energy. A measurement of angular momentum involves measurements of both axes, the semi-major and the semi-minor. Angular momentum assessments concern both the semi-major and the semi-minor axes.

As was mentioned earlier, for an ellipse, the "x" axis measures its greatest length while the "y" axis measures its greatest width. An eccentricity of .000000 makes an ellipse into a perfect circle, and the semi-major axis and the semi-minor axis are equal in length. Today, Venus has an orbit eccentricity of only .006783 and its orbit is nearly a perfect circle. Its modern perihelion is 67,237,928 miles from the Sun (the semi-major axis) and its semi-minor axis is only 1,546 miles shorter, at 67,236,382 miles.

Halley's Comet on the other hand illustrates an orbit of high eccentricity, almost .75. Tables XII and XIII show that the eccentricity of the Martian orbit in the Catastrophic Era was close to .561. In the modern age, the Martian eccentricity is only .093.

As it is with the conservation energy, so it is with the conservation of angular momentum. Angular momentum may be exchanged with another planet during a close flyby. But unlike energy, angular momentum constantly is being exchanged also at great distances.

Still, what one planet gains in angular momentum, that same amount the exchanging planet must lose. Like energy, angular momentum neither appears nor disappears by magic, nor can it disappear (or appear) by the stroke of the pen of some wishful philosopher or hopeful space cadet.

The present problem, or issue, is how the Martian orbit changed from what it was in the Catastrophic Era to the modern age, or, how the orbit of Mars rounded out. If energy exchanges were the only issue, there are thousands of theoretical solutions that would satisfy such an energy exchange. Similarly, if angular momentum were the only issue, there are thousands of possible solutions.

Ideally, there is only one solution that will simultaneously satisfy both exchanges. Assumptions are made about the Catastrophic Era orbits that can be assumed from ancient resonance's but cannot be verified. Therefore the solution in Tables XI, XII and XIII (chapters 9-10) are at best an approximate solution. Table XIII may contain a more exact solution, but at the least it is very very close. This solution is hard to popularize for the general reader; it is in the arena of celestial mechanics and orbital analysts for examination and assessment.

If a theory of past planet warfare could not be demonstrated mathematically how Mars shifted from its catastrophic era orbit to its modern orbit, it is a flawed and a speculative theory.

Herein is the logic of the conversion of the orbit of Mars from its old orbit in the Catastrophic Era into its modern condition. While the participation of the Sun, the Earth-Moon system and Mars seem, basic, in reality the participation of both of Venus and Jupiter also are essential. (Any model involving only the Sun, the Earth-Moon system and Mars will inevitably fail).

This model as described in Tables XI, XII and XIII requires that on or about January 24, 701 B.C.E., plus or minus 1 day, this trio of planets (Mars, Venus and the Earth) were poised for battle. Cosmic battle. Energy exchanges. Angular momentum exchanges. Eccentricity changes. Altered orbits. First came the last Ares-Pallas Athene planetary polka.

The Venus versus Mars battle preceded an Earth versus Mars battle by some 56 days, plus or minus one. Energy shifted in equal amounts; what energy and momentum Venus lost Mars gained in equal amounts. Alterations of orbits occurred. Spin axes precessed. Crusts bulged due to heavy subcrustal tides. Planetary magnetic field reunions were held, briefly, just for hours, somewhat like class reunions.

The action, when it came, was fast, ferocious, fierce, furious, and in year 701 B.C.E. (It was final unless new planets are to be delivered to the Sun's planetary domain in some future catastrophic scene). When it was all over, the modern era for the orbits of Venus, Mars and the Earth-Moon system had arrived. And with it came Solar System gradualism, a system now only 2701 years old in 1997.

The orbit of Mars rounded out. No more celestial waltzes with the Earth. No more passionate polkas with Venus. No more fast and furious flings. No more precessing gyroscopes in the cosmos. No more electrical-celestial trysts in space between clinging planetary magnetic fields. (There is no more chaos in the foreseeable future).

The Angular Momentum Exchange

With the Energy Exchange, the total energy budgets for the four planets involved always had to balance at the beginning, at the end and at the finish of each of the four stages. As it was in the energy exchange, so it also was with angular momentum budgets.

Energy shifts and momentum shifts also involve sudden tides in the oceans, crustal earthquakes, magma tides, elastic crustal bending, volcanic eruptions, planetary magnetic field reunions, spin axis shifts and changes in satellite orbits. But all these eight conditions are insignificant if an analysis of energy and momentum budgets is carried to just six decimals, as is the case here. To carry it farther to ten or twelve decimals and include these factors would be to present a false sense of extreme accuracy.

The equation for angular momentum for each planet is as follows:

2 pi x mass x sq. root (semi-major axis x [1 - ecc.2])

Table XII - The Angular Momentum Exchange

January 24, 701 B.C.E.

Planet	Mass	1/2 "x" axis	Orbital Eccentricity	Angular Momentum
Old Venus	0.814988	0.726143	0.006000	4.363531
Old Earth	1.012303	0.993367	0.015000	6.338644
Old Mars	0.107447	1.576872	0.560966	0.701808
Old Jupiter	317.799873	5.202804	0.047829	4549.409178

Angular Momentum Total - January 24, 701 B.C.E. 4560.813161

AFTER THE FINAL FLING BETWEEN MARS AND VENUS
Approx. Jan. 25, 701 B.C.E.
Mars gained .008514 angular momentum units
Venus lost .008514 angular momentum units

End of February 701 B.C.E.

Planet	Mass	1/2 "x" axis	Orbital Eccentricity	Angular Momentum
New Venus	0.814988	0.723332	0.006783	4.355017
Old Earth	1.012303	0.993367	0.015000	6.336345
Interim Mars	0.107447	1.685225	0.585742	0.710322
Old Jupiter	317.799873	5.202804	0.047829	4549.409178

Angular Momentum Total - February 1, 701 B.C.E. 4560.813161

AFTER THE FINAL WALTZ BETWEEN MARS AND THE EARTH
March 20-21, 701 B.C.E.
The Earth-Moon system gained .020954 ang. mom. units
Mars lost .020954 ang. mom. units

End of March 701 B.C.E.

Planet	Mass	1/2 "x" axis	Orbital Eccentricity	Angular Momentum
New Venus	0.814988	0.723332	0.006783	4.355017
New Earth	1.012303	1.000000	0.016718	6.359598
Interim Mars	0.107447	1.523691	0.561858	0.689368
Old Jupiter	317.799873	5.202804	0.047829	4549.409178

Angular Momentum Total - 700 B.C.E. 4560.813161

THE FINAL SLOWER TWEAKING OF MARS BY JUPITER
March 22, 701 B.C.E. for the next 30 years or more
Mars gained .140331 angular momentum units
Jupiter lost .140331 angular momentum units

Planet	Mass	1/2 "x" axis	Orbital Eccentricity	Angular Momentum
New Venus	0.814988	0.723332	0.006783	4.355017
New Earth	1.012303	1.000000	0.016718	6.359598
Mars	0.107447	1.523692	0.093387	0.829699
Jupiter	317.799873	5.202804	0.048468	4549.268847

Angular Momentum Total - Modern Age 4560.813161

Tables XI and XII are both condensations of the orbital changes of the planets statistically. Table XIII is an elaborate table, including 24 categories of data.

Figure 22 portrays the positions and the orbits of Venus and Mars on the eve of the Mars-Venus Polka in January, 701 B.C.E. Figure 25 in chapter 11 describes the positions and the orbits of Mars and the Earth as they were poised for the Final Flyby.

The Mars-Venus Fling and the Mars-Earth Final Flyby were spasms of catastrophism (and crustal diastrophism). Each came, occurred with violent developments and were over in 24 hours. The last stage, the interaction between Mars and Jupiter was different. It was a slow affair, at least 30 years long and was conducted across great distances. It produced a radical change in the Martian eccentricity, as the orbit of Mars rounded out.

Figure 22 - Positions of Mars, Venus, and Earth on the Eve of the Mars-Venus Polka, Approximately Jan 23, 701 B.C.E.

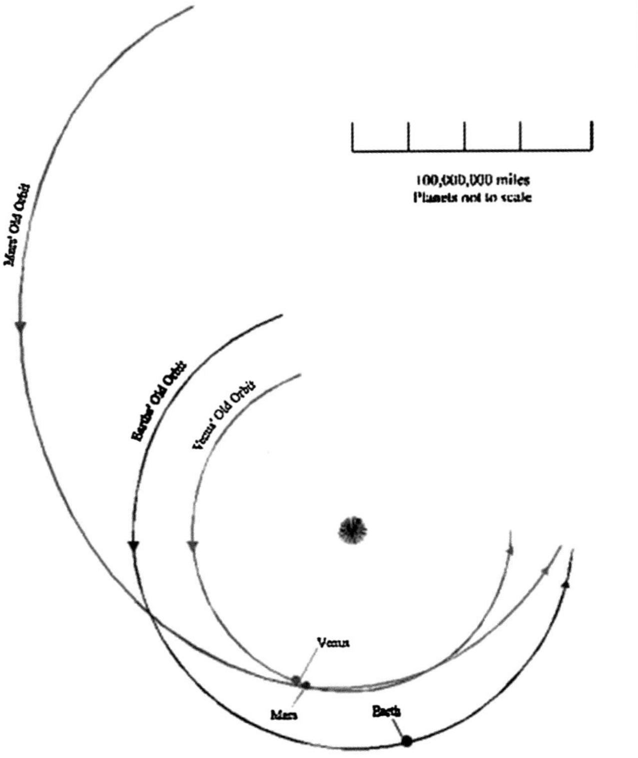

Positions of Mars, Venus, and Earth on the Eve of the Mars-Venus Polka, Approximately Jan 23, 701 B.C.E.

Figure 23 - Geometry and Distances of The Last Mars-Venus Flyby

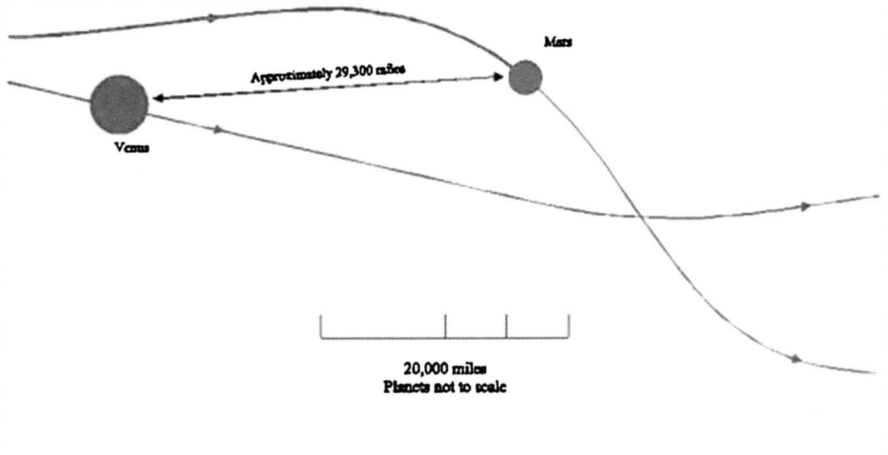

Geometry and Distances of The Last Mars-Venus Flyby

This model needs to be expanded into greater detail for the expert in mathematics, physics and astronomy. It contains 24 categories of data to enlarge the picture of these two spasms of catastrophism, plus the slower influence of Jupiter from afar. This is how Mars departed from its ancient 6:1 resonance with Jupiter and acquired a radically. changed orbital eccentricity.

The Changes To Four Planetary Orbits

Formulas And Constants

AM=Angular Momentum
SMn=Semi-Minor
SMj=Semi-Major
AM=2PI*Mass*sqr(SMj*(1-e^2))
energy=(2PI^2*Mass)/SMj
SMj=f/(1-e)
e=(SMj-f)/SMj
f=SMj*(1-e)
Earth mass=5.9742 * 1024 Kg
1 mile=1.609344 Km
1 au=149597870 Km
Earth period now=365.256365 days
Period for a planet=square-root((SMj3*Earth's period2)/Earth's SMj3))

Table XIII - Detailed Changes In Planetary Orbits In The 8th Century B.C.E. - 24 Categories of Astronomical Data of the Model

PERIOD	MASS		Semi Major	Axis	Semi Minor	Axis
PLANET	Kg 10^{24}	Earth base	miles	au	miles	au
PRE-Venus flyby						
Venus	4.8689	0.814988	67,500,395	0.726156	67,499,180	0.726143
Earth-Moon	6.0477	1.012303	92,339,242	0.993367	92,328,853	0.993255
Mars	0.64191	0.107447	146,579,410	1.576872	121,344,158	1.305396
Jupiter	1898.6	317.799873	483,630,834	5.202804	483,077,339	5.196850
POST-Venus	PRE-Earth flyby					
Venus	4.8689	0.814988	67,237,928	0.723332	67,236,382	0.723316

Earth-Moon	6.0477	1.012303	92,339,242	0.993367	92,328,853	0.993255
Mars	0.64191	0.107447	156,651,477	1.685225	126,965,676	1.365871
Jupiter	1898.6	317.799873	483,630,834	5.202804	483,077,339	5.196850
POST-Earth flyby	PRE-Modern					
Venus	4.8689	0.814988	67,237,928	0.723332	67,236,382	0.723316
Earth-Moon	6.0477	1.012303	92,955,807	1.000000	92,942,817	0.999860
Mars	0.64191	0.107447	141,635,971	1.523691	117,166,093	1.260449
Jupiter	1898.6	317.799873	483,630,834	5.202804	483,077,339	5.196850
MODERN						
Venus	4.8689	0.814988	67,237,928	0.723332	67,236,382	0.723316
Earth-Moon	6.0477	1.012303	92,955,807	1.000000	92,942,817	0.999860
Mars	0.64191	0.107447	141,635,973	1.523692	141,017,014	1.517033
Jupiter	1898.6	317.799873	483,630,834	5.202804	483,062,438	5.196689

PERIOD	Perihelion		Aphelion	
PLANET	miles	au	miles	au
PRE-Venus	flyby			
Venus	67,095,393	0.721799	67,905,398	0.730513
Earth-Moon	90,954,153	0.978467	93,724,331	1.008268
Mars	64,353,387	0.692301	228,805,433	2.461443
Jupiter	460,499,288	4.953959	506,762,381	5.451648
POST-Venus	PRE-Earth flyby			
Venus	66,781,880	0.718426	67,693,976	0.728238
Earth-Moon	90,954,153	0.978467	93,724,331	1.008268
Mars	64,894,175	0.698119	248,408,779	2.672332
Jupiter	460,499,288	4.953959	506,762,381	5.451648
POST-Earth flyby	PRE-Modern			
Venus	66,781,880	0.718426	67,693,976	0.728238

Earth-Moon	91,401,818	0.983283	94,509,796	1.016718
Mars	62,056,734	0.667594	221,215,207	2.379789
Jupiter	460,499,288	4.953959	506,762,381	5.451648
MODERN				
Venus	66,781,880	0.718426	67,693,976	0.728238
Earth-Moon	91,401,818	0.983283	94,509,796	1.016718
Mars	128,409,085	1.381399	154,862,861	1.665984
Jupiter	460,190,167	4.950634	507,071,502	5.454974

PERIOD	Eccentricity	PERIOD	PERIOD	RATIO
PLANET		Modern days	to Earth	to Jupiter
PRE-Venus	flyby			
Venus	0.006000	226.017716	0.625000	0.052142
Earth-Moon	0.015000	361.628345	1.000000	0.083427
Mars	0.560966	723.256691	2.000000	0.166855
Jupiter	0.047829	4334.649423	11.986476	1.000000
POST-Venus	PRE-Earth flyby			
Venus	0.006783	224.700734	0.621358	0.051838
Earth-Moon	0.015000	361.628345	1.000000	0.083427
Mars	0.585742	799.069854	2.209644	0.184345
Jupiter	0.047829	4334.649423	11.986476	1.000000
POST-Earth flyby	PRE-Modern			
Venus	0.006783	224.700734	0.615186	0.051838
Earth-Moon	0.016718	365.256365	1.000000	0.084264
Mars	0.561858	686.978823	1.880813	0.158485
Jupiter	0.047829	4334.649423	11.867417	1.000000
MODERN				
Venus	0.006783	224.700734	0.615186	0.051838
Earth-Moon	0.016718	365.256365	1.000000	0.084264
Mars	0.093387	686.978839	1.880813	0.158485
Jupiter	0.048468	4334.649423	11.867417	1.000000

PERIOD	Energy	Angular	Semi Major	Semi Minor	Perihelion
PLANET		Momentum	Change	Change	Change
PRE-Venus	flyby				
Venus	-22.153944	4.363531	262,467	262,799	313,513
Earth-Moon	-20.115482	6.338644	-616,565	-613,963	-447,665
Mars	-1.345017	0.701808	4,943,437	-19,672,856	-64,055,698
Jupiter	-1205.718718	4549.409178	0	14,901	309,121
Total	-1249.333160	4560.813161			
POST-Venus	PRE-Earth flyby				
Venus	-22.240423	4.355017	-262,467	-262,799	-313,513
Earth-Moon	-20.115482	6.338644	0	0	0
Mars	1.258537	0.710322	10,072,067	5,621,518	518 540,788
Jupiter	-1205.718718	4549.409178	0	0	0
Total	1249.333160	4560.813161			
POST-Earth flyby	PRE-Modern				
Venus	-22.240423	4.355017	0	0	0
Earth-Moon	-19.982058	6.359598	616,565	613,963	447,665
Mars	-1.391961	0.689368	-15,015,507	-9,799,583	-2,837,441
Jupiter	-1205.718718	4549.409178	0	0	0
Total	1249.333160	4560.813161			
MODERN					
Venus	-22.240423	4.355017	0	0	0
Earth-Moon	-19.982058	6.359598	0	0	0
Mars	1.391961	0.829699	2	23,850,921	66,352,351
Jupiter	-1205.718718	4549.268847	0	-14,901	-309,121
Total	1249.333160	4560.813161			

PERIOD	Aphelion	Eccentricity	Period	Energy	AM
PLANET	Change	Change	Change	Change	Change
PRE-Venus	flyby				
Venus	211,422	-0.000783	1.316982	0.086479	0.008514
Earth-Moon	-785,465	-0.001717	-3.628020	-0.133424	-0.020954
Mars	73,942,572	0.467579	36.277851	0.046944	-0.127891
Jupiter	-309,121	-0.000639	0.000000	0.000000	0.140331
			Total	0.000000	0.000000
POST-Venus	PRE-Earth flyby				
Venus	-211,422	0.000783	-1.316982	-0.086479	-0.008514
Earth-Moon	0	0.000000	0.000000	0.000000	0.000000
Mars	19,603,346	0.024776	75.813163	0.086479	0.008514
Jupiter	0	0.000000	0.000000	0.000000	0.000000
			Total	0.000000	0.000000
POST-Earth	flyby PRE-Modern				
Venus	0	0.000000	0.000000	0.000000	0.000000
Earth-Moon	785,465	0.001717	3.628020	0.133424	0.020954
Mars	-27,193,572	-0.023884	-112.091031	-0.133424	-0.020954
Jupiter	0	0.000000	0.000000	0.000000	0.000000
			Total	0.000000	0.000000
MODERN					
Venus	0	0.000000	0.000000	0.000000	0.000000
Earth-Moon	0	0.000000	0.000000	0.000000	0.000000
Mars	-66,352,346	-0.468471	0.000016	0.000000	0.140331
Jupiter	309,121	0.000639	0.000000	0.000000	-0.140331
			Total	0.000000	0.000000

VENUS. Mars made an inside flyby of Venus at about 29,300 miles at its closest. Venus yanked Mars outward; Mars yanked Venus inward. The orbit of Venus shrunk; its semi-major axis was lessened by some 262,467 miles, or 0.389%. Its period was reduced by 1.317 days, or 0.583%. In its new orbit Venus began to receive 0.6% more solar radiation than formerly. The hot planet became slightly hotter.

The amount of energy that Venus lost and Mars gained during its last Mars polka was .086479 energy units. The amount of energy the Earth gained during its last Mars waltz was .133424 energy units, or 154.285% as much as Venus lost. If the closest distance during the last Mars-Earth waltz was 27,000 miles, it calculates that the closest distance of the last Mars-Venus polka was approximately 29,300 miles.

The new period for Venus became 224.7 days, down from its ancient 226.017716 days. Its year shortened by 0.583%.

JUPITER. Jupiter's distance from the Sun, its semi-major axis did not change. Its eccentricity, or oblateness, increased from .047829 to .048468. This is a increase of 1.336%, or one part in 74.85.

Jupiter is 2,957 times as massive as little Mars. When Mars rounded out, 100.191% of its decrease in eccentricity (rounding out) was caused by Jupiter, 5.108% was caused by the Earth and -5.299% was caused by Venus. Mars left its former 6:1 resonance with Jupiter. This change, and Jupiter's huge remote mass, is what caused Jove's energy to remain unchanged.

MARS. During these three cosmic scenes, the length of the semi-major axis of little Mars decreased, but only by 4,943,437 miles, or 3.37%. However, Venus caused its x-axis to increase by 6.87%. But the Earth decreased the red planet's distance from the Sun by an immense 10.24%. This is one reflection of the passionate, energetic nature of the final Mars-Earth waltz.

During this three-act drama, the period of Mars was decreased from 723+ days to 686.98, a decrease of over 36 days. With the changes in eccentricity, the new Mars never comes closer to the Earth than 25.6 million miles.

In the old catastrophic scenario, on this occasion Mars came within an estimated 27,000 miles of the Earth, planet center to center. Some, possibly 5% of the ancient flybys were even closer (but they were of the opposite or inside geometry). The Noah's Flood Flyby, - 2484 was such an occasion; the Long Day of Joshua Flyby, - 1404, was another.

THE EARTH. The Earth's last waltz with Mars was close and passionate. Their planetary magnetic fields united in space for ten or fifteen hours, while their gravitys "fought" each other's. Then, after a brief magnetic tryst in space, that briefly unified field was ripped apart, for the last time.

In the Catastrophic Era, Mars flybys charged and recharged the Earth's geomagnetic field, which decayed slowly between flybys. But after the last flyby, the Earth's geomagnetic field strength has began to decay at a rate of a half life every 1,350 years, with no recharging.

The Earth's average distance to the Sun, its semi-major axis, was increased by 616,000+

miles, or 0.668% In the new era, the Earth receives almost 1.5% less solar radiation, and it has become slightly colder.

The Earth's spin axis decreased an estimated 0.452% from 1442 minutes to 1436 minutes (sidereal). The Earth's period increased from 360 old days to 365.256 new days. About 30.97% of the increase in day count per year (5.256 days) was due to spin rate increase. And 69.03% was from orbital expansion.

Thus it was that after this last waltz, dozens of ancient 360-day calendars in various civilizations all simultaneously became unsatisfactory and obsolete. A historical analysis reports that all ancient calendar reorganizations followed the year 701 B.C.E. by various lapses of time.

As was mentioned earlier, the ancients had an approximate 360-day year, but it wasn't exact. It was a measure of rotations in the Earth's old year --- the Earth's old spin rate --- versus the Earth's old orbit. The ancient Egyptians had two calendars. One was the 360-day per year calendar. The other, more precise and apparently technically better, was their Venus calendar.

Tables XI, XI and XIII incorporate this information, and provide enough information to calculate the loss of period, of energy and of angular momentum for old Venus in its last Mars-Venus Polka. Compared to the Earth's old year, 1.000, the period of the old Venus orbit was .62500, or exactly five-eighths of the Earth's period. The current comparison is Venus at .61519 of the Earth's period.

THE MOON. Ancient civilizations on at least four continents had the lunar period at 30 days. The modern sidereal period is 29.53 days. This means that when Mars passed between the Earth and the Moon (for the first time in 701 B.C.E.), the red planet was some eight times closer to the Earth than to the Moon (at its closest distance).

In so doing, not only did Mars pull the Earth out some 616,000 miles. The red planet also pulled the Moon inward. The inward change is estimated at 2,350 miles, or about 1% in distance from an old semi-major axis of 241,200 miles Now it is 238,850 miles.

On Scientific Discoveries

The problem of solving simultaneously energy and angular momentum exchanges simultaneously has consumed more than some five years. It has involved numerous conversations involving Ronald R. Hatch, Michael T. Lute, Samuel R. Windsor and Donald W. Patten. The conclusions are only as good as are the assumptions that have been made. .

It is shown that the orbit of Venus was in 8:5 resonance with the Earth, when it served for long term Egyptian calendars. In this case, since the Earth was in 2:1 resonance with

Mars, in 5:8 orbital resonance with Venus, in 12:1 resonance with Jupiter, and as it happens, in 85:1 resonance with Uranus.

The ancients, including author of the Book of Job, discussed an ancient harmony of the spheres in the heavens. Perhaps this work in Chapters 9 and 10 has explained what that ancient harmony was.

Many times in the history of science, there have been feelings of elation in making scientific discoveries after a long, grueling and eventually successful search. It was so with scientists in the past, such as Johannes Kepler, Louis Pasteur, Claude Bernard, D. F. Frazer-Harris, James Clerk-Maxwell and many others. With respect to the work of R.E.D. Clark of England, the words of Kepler, Pasteur, Bernard, Frazer-Harris and R. E. D. Clark are cited.

> *The intense pleasure I have received from this discovery can never be told in words.* - Johannes Kepler [n3]
>
> *When you have at last arrived at certainty, your joy is one of the greatest than can ever be felt by a human soul.* - Louis Pasteur
>
> *The joy of discovery is certainly the loveliest that the mind of man can ever feel.* - Claude Bernard
>
> *The joy of scientific discovery is one of the most exalted human emotions.* - D. F. Frazer-Harris
>
> *We have asked questions both about the beginnings of science and about the way we think. Let us now consider scientific ideas as they take shape in the minds of discoverers. First, let us discuss some case histories of "hunches" - flashes of creative intuition which bring new ideas into our world.* - R. E. D. Clark [n4]

Conclusion

Understanding energy shifts is one essential in comprehending the increase of the Earth's orbit from an ancient 360 days to the modern 365.256 days. It is also an essential in understanding the shift down in the orbit of Mars from 720 old days to 686.98 modern days. Additionally, it is of some interest to comprehend the shift down in the Moon's orbit from the ancient 30 days to 29.53 modern days. .

Understanding especially ANGULAR MOMENTUM shifts are essential in comprehending the rounding out of the Martian orbit. The Earth helped marginally, but Jupiter's influence was by far the dominant factor. The rounding out of the Martian orbit is what ended the catastrophic era. It happened at the time of the vernal equinox of 701

B.C.E., the night of March 20-21, the final celestial Passover. (Annual religious Passovers have occurred in Judaism annually ever since the time of Moses, when the Hebrews fled and achieved freedom from Egyptian bondage on a Passover night.).

Tables XI and XII are condensations, and in brief illustrate statistically what happened to the orbits of Venus, the Earth, Mars and Jupiter at the end of the Catastrophic Era. Table XIII provides a wider range of information, 24 categories rich in details. This table is addressed to experts in the Second Law of Thermodynamics.

Story 34 concerns the aging of the Solar System. As it was in the case of the asteroids, so it has been with the planets Mars, the Earth and Jupiter. When orbiting planets are in orbit resonances, IT IS A SIGN OF THE YOUTHFULNESS OF THAT SYSTEM.

In experiencing catastrophes and their perturbations, planets, like asteroids, have shifted increasingly away from ancient resonances. Order has decreased. The shift into non-resonant orbits is AN INDICATION OF AGING HAVING OCCURRED, AND IT IS INDICATIVE OF A RECENTLY FORMED SOLAR SYSTEM. The Solar System thus has passed from infancy into childhood. Note the many levels of evidence cited in "The Recent Organization of the Solar System". [n5]

The number of Mars-Venus flybys is unknown, but the surface of Venus does have a battered physical geography. That battered geography of its crust is not a topic in this volume. It might be that Venus experienced as many Mars-Venus Wars as did the Earth. It might be that they were, on average, equally severe. The consensus among geologists is that there seems to have been 170 paleomagnetic polarity reversals - 170 Mars flybys.

Story 35 is that the orbit of Mars converted from one of high eccentricity, ONE OF .561, to a radically changed to orbit, one moderately low in eccentricity, one that rounded out to .093. This occurred while energy diminished only 3.49% in the process.

It cannot be proved, but it is probable that in the events of -701, the orbits of Venus and the Earth each experienced a modest or minute increase in orbit eccentricity. Clearly, Venus experienced a decrease in orbit size whereas the Earth, experiencing an energetic outside flyby by Mars, acquired a somewhat enlarged orbit.

How many scars of Mars is needed to establish it as the cosmic felon that destroyed Astra ... and assaulted Venus and the Earth. The red planet has one badly battered, cratered up hemisphere, and one remarkably serene hemisphere with few (only 7%) of the Martian craters.

However, the Serene Hemisphere of Mars has one massive rift system, two huge bulges and several enormous volcanoes. Its badly bombarded cratered hemisphere contain 93% of all of its craters. Its ancient orbit indicates it went out to where the asteroids now roam, and that it was the felon causing the death by fragmentation of a former planet. Only 5,000 fragments (asteroids) remain. The poxed Deimos and Phobos are two of them.

The poxed asteroid Gaspra, speckled with craterlets, is also a scar of Mars, and possibly was part of its former ring system. The craterlets on Deimos and Phobos also are scars of Mars. So are its dry river beds and its ancient cometary tail, two more scars of Mars, one scar being physical, one literary in nature.

The twin spin axis tilts of Mars and the Earth also are scars of the Mars-Earth Wars. Thus, there are scars of Mars from within the crust of Mars, its spin axis shifts and its bulges. The scars of Mars are on its surface, its craters, its rifts, its volcanoes. And scars are seen beyond its surface, craterlets on the surfaces of its two tiny satellites. The physical geography of Mars and its satellites is fascinating (to say the least). And instructive.

Story 36 in our skyscraper of catastrophic cosmology is that calculations now have been made SHOWING HOW AND WHEN MARS SHIFTED INTO ITS MODERN ORBIT. It was not achieved billions of years ago, not millions of years past, not even hundreds of thousands of year ago. It was achieved 2,700 years ago. All gradualists please take note.

Some astronomers have supposed that it is impossible to reconcile energy shifts and simultaneous angular momentum shifts to move the Earth from a 360-day to a 365+-day orbit and Mars from a 720-day orbit to one of 687 days. Included is a shift in eccentricity from .561 to .093. Three such skeptics have assumed, mistakenly, they were facing a three-body problem (Earth, Mars and Sun). In fact, they were facing a five-body issue, the Sun, Venus, the Earth-Moon system, Mars and Jupiter.

The energy totals agree at the beginning and at the end, plus at the end of each stage. So do the angular momentum totals. Mars went out into the asteroid belt to destroy the former planet Astra. The red planet came inside the orbit of Venus and tortured that planet occasionally. The perplexing problem is solved, but it mandates a paradigm shift in thinking.

Simultaneously, this model describes (a) the rounding out of the ancient orbit of Mars, (b) the slight shortening of the Moon's period, and (c) the increase of the Earth's period from about 360 old days to 365.256365 modern days.

HOMER. Reportedly a blind bard, yet Homer understood something about Ares assaults and some aspects of planetary catastrophism, some 2,850 years ago. He described Ares assault scenes early on in The Iliad. That writing and one other, the Odyssey, has made Homer famous across the centuries, and even across three millenniums.

If blind old Homer could, in part, understand Ares assaults and planetary catastrophism, then with all of their modern resources, there is no reason late 20th century astronomers and geologists cannot understand this also. This is despite the fact that they themselves, as young students, were victims of 150 years of mistaken anticlerical gradualist dogmas.

The challenge is for 21st century scholardom to rise out of this 18th and 19th century stupor, and from its 20th century cosmological malaise.

Story 37 is the Barringer Crater which is in Northern Arizona, diameter 4,150 feet, depth 570 feet. It apparently was a result of a Mars asteroid hitting Northern Arizona. It was in an era of which the Hopi Indians still retain memory of the occasion. This Arizona asteroid accident likely had been part of the former ring system of Mars, companions of Deimos and Phobos.

The fall of an asteroidal body and the creation of the Barringer Crater is in Hopi ancient traditions and stories. The ancient Hopi were the mysterious Anasazi. Indians have a long memory, and this is an illustration. The Barringer Crater is also one of the Scars of Mars. Ironically, this particular scar from Mars is located only 45 miles distant from the Percival Lowell observatory at Flagstaff. Lowell was the most famous advocate of the 20th century "men on Mars".

The reader now is 74% of the way to an integrated system, or paradigm of ancient history with ancient geological upheavals and astronomical upheavals..

As it is with the view at the Space Needle Restaurant in Seattle, or climbing Mount Rainier, so it is with this model. As elevation increases, the more spectacular and majestic the celestial scenery becomes.

CHAPTER 11

The First Nine Clues

Grab the enemy by the nose and kick him in the pants.
George S. Patton Jr.

Introduction

This model of planetary catastrophism includes first, Mars making a flyby of Venus, an energetic inside flyby of Venus at the 76th longitude. It was on January 24, +/- 1 day, 701 B.C.E. This was where the orbits of Mars and Venus crossed as Mars approached its old perihelion in that era.

Next and last, Mars is modeled as having made an even more energetic flyby, this time a unique outside flyby, of the Earth at longitude 179 or 180. That is where the orbit of Mars crossed the Earth's orbit for the last time. This second energetic flyby, 55, 56 or 57 days later, occurred on the night (Near East Time) of March 20-21, 701 B.C.E.

This model, for the year 701 B.C.E., is one of double-barreled catastrophism. Evidence might be found either contradicting or supporting this model. For instance, the position of the Moon, modeled as being at full moon on the night of March 20-21, 701 B.C.E., can be retro-calculated. Does the Moon retro-calculate back, using a 365.256-day year, to a full moon on March 20-21, 701 B.C.E.?

And what about retro-calculating the positions for Mars, Venus, and Jupiter? Will their retro-calculations give testimony for or against this model?

If such retro-calculations were correctly made, and were in disagreement with this model, this model would be flawed. If only some of the clues were in agreement, the model would be partly flawed. If retro-calculations of all three planets and the Moon are found to be in agreement with the model, it will be a confirmation. It is comparable to a murder mystery, where a detective listens to the alibis of the four suspects of the crime. Next he proceeds to verify the veracity of the alibis.

Chapter 11 includes nine of twelve categories of clues; each is a test for the validity of the model. Chapter 12 contains clues 10 to 12.

Four clues, or tests, for this model are retro-calculations back to March 21, 701 B.C.E. for the positions of the key planets, the Moon's position, and the positions of Venus, Mars and Jupiter. These are discussed in clue 1 (the Moon), clue 2 (Venus), clue 4 (Mars) and clue 6 (Jupiter).

The third clue is the semi-major axis, and the location of the perihelion of Venus. Does its position also contain some kind of a clue of Mars-Venus catastrophism? Evidence of violence on the surface of Venus has been found, but the specifics have not been cited herein. If there was violence to its crust, perhaps there was violence to its orbit as well? Perhaps there is evidence thereof.

The fifth clue is the modern orientation orbit of Jupiter and its semi-major axis. It is modeled that Jupiter's orbital period and the Earth's were in a 12:1 resonance, and the Martian orbit was in a 6:1 resonance with Jupiter. Are there vestiges pointing to such a former relationship?

A seventh clue involves the distribution of asteroids in the asteroid belt. There is a certain feature, a gap in the distribution of asteroids at the 2:1 resonance location. It may be instructive. It concerns asteroids that have been perturbed out of resonance, and a characteristic of them. Does the modern orbit of Mars contain a parallel characteristic?

It is recalled that the Earth is modeled as formerly having been in 2:1 orbital resonance with Mars. But no longer. Is there anything common about asteroids that have left 2:1 resonance and planets that have left 2:1 orbital resonance? These are technical questions, but such is the nature of little clues.

The eighth clue involves the twin spin axis tilts of the Earth and Mars. Both tilts are close to 22.5°. This was discussed in chapter 8. Is there any way other than alternating, reciprocal planetary catastrophism for those two spin axes to have twin tilts?

The ninth clue involves the spin axis of the Earth. It responds to lunar and solar tides, and in its response, the spin axis follows a great circle in the northern heavens, a circle which takes 25,800 years to complete. Today, Polaris is the North Star, where the spin axis points. It will point there again in 25,800 years. Where did the spin axis point during the Catastrophic Era according to ancient accounts?

Ursa Minoris was the pole star when the Final Flyby occurred, 2,700 years ago. It is "Kochab" in Arabic, which means "The Pole Star". Has luni-solar tides, and consequent spin axis precession been ongoing forever in 25,800-year cycles? 1,550 of them would cover the last four billion years? Or is there evidence that planetary catastrophism has disrupted and reorganized luni-solar precession?

The tenth clue is the 360-day calendars, calendars so popular with the ancients on five continents. And the tenth clue cites their 360-degree circles, also popular with ancient mathematicians and surveyors on three continents.

If the Earth has always had a 365.256-day year, why were 360-day calendars used on every continent and 360-degree circles were also used in China, the Middle East and the Near East? Perhaps calendars, like other time pieces, can tell a story and if so, what is their message? Is it that the ancient mathematicians and calendar-makers couldn't count?

The eleventh clue is the mystery clue.

The twelfth clue involves the development of the English language from a variety of ancient languages, including Anglo Saxon, Scandanavian and Germanic languages, French, Latin, Greek and a smattering of other ancient sources. Their words for Mars included words such as Ares, Baal, Bel, Horus, Indra, Mars, Nergal and the Chinese dragon star.

Perhaps the ancient Greek words for little Deimos and Phobos also have left their imprints in modern English.

How many words in modern English are derived from various ancient words for Mars? 100 words, 200, 300, 400 or even 500 words. What kind of thoughts or experiences do those Mars-related words suggest? "March", for instance. Or "dis-aster" (aster = planet). Or cat-astro-phe. "Cata" means thrown downward from the heavens.

What kind of words and thoughts would these modern words in 20th century English tend to convey? Are these words linguistic vestiges of the Mars-Earth Wars? Are they linguistic scars of Mars?

Clue # 1 - The Moon And The Final Flyby Of Mars

This model states that the Moon's position was at "full moon" during the Final Flyby on the night of March 20, 701 B.C.E. It was the Hebrew Passover, and the Moon always was full on Passover nights. On the Hebrew Nisan calendar, it was Friday, the 13th of Nisan, our night of March 20. According to ancient Hebrew literature, each of the twelve months had 30 days. The night of the 7th was always a new moon, and the night of each "Passover" was always during the full moon in the month of Nisan.

This catastrophic flyby was on the anniversary of a long series of anniversaries of "unlucky" Friday the 13ths. In the modern era, the myth of bad experiences of this particular night, Friday the 13th, comes down to our age through Hebrew folklore. It was in an era when cosmic developments of that night really were a bad experience.

The Irish are not to be easily outdone. Another theme, the Halloween theme, a late

October flyby theme, fairly similar, has been passed down from the ancient Celtic druids of Ireland and Great Britain. This type of flyby coincided with the October case flybys, October 24.

Recently the phase of the Moon was again at full on December 6, 1995. This is 2,700 years lacking a few months of the Final Flyby. According to the model, prior to the Final Mars Flyby, the Moon's orbit was in 12:1 resonance with the Earth's old orbit, and 30 days comprised an old month as did 360 days comprise a year.

According to the model, the Moon was about 242,000 miles from the Earth. The period of the Moon suddenly contracted to 29.53059 days and its average distance became 238,860 miles.

Its new orbit, and orbital period were established, according to the model, on the day immediately after the Mars Flyby, on March 22, 701, as Mars swept through the Earth-Moon system, between the two. The red planet came about eight times closer to the Earth, at 27,000 miles, than to the Moon, at 215,000 miles.

In this close flyby, the Fire Star, Ares, pulled the Earth-Moon system out over 690,000 miles. Simultaneously Ares, bane of mortals, pulled the Moon inward, closer to the Earth, by about 3,000 miles.

The Moon's position can be retro-calculated backward to March 20, 701 B.C.E. When doing this, it is important to know whether the year for "zero" between B.C.E. and AD is counted. Some do it one way; some the other. Usually, it is not counted, as is the case here. Further, in calculations, it is necessary to include an error, a four-year error, made by medieval monks.

THE DIONYSIUS EXIGUUS FACTOR. Dionysius Exiguus was a medieval monk, who was given the task of resolving a calendar dispute as to the proper date for Easter. Later research revealed Dionysius had missed four years in assessing the year of Christ's birth. Many centuries before this error was identified, his dating system for history had come to be accepted.

His error was not revised, in order to minimize confusion. By the time his error was realized, the sequencing of historical dates for the Roman Empire and for early Christianity had long ago become too widely accepted. So Dionysius' dates were kept, and mankind was left with a quixotic system. The accepted system cites that Christ was born in 4 B.C.E., seemingly an impossibility. Inclusion of the Dionysius factor is essential in any careful analysis.

The calculations for retro-calculating the Moon's position are as follows.

1. March 21, 701 B.C.E. to March 20 1 B.C.E. 700 years

2. March 21, 1 B.C. to March 20, 1 AD 1 year

3. March 21, 1 AD to March 20, 1995 1994 years

4. March 21, 1995 to Dec. 6, 1995 260 days

5. Add 4 years for Dionysius Exiguus 4 years

Total 2,699 years 260 days

The time lapse 2,699 years and 260 days is also 2699.7118 years, which also is 986,087 days. The Moon's modern period is 29.53059 (synodic). This day count, 986,087 days, when divided by 29.53059, determines where the Moon was on the night of the Final Waltz. Between these dates, both full moons according to the model, the Moon has made 33,392.05 orbits. The full moon enjoyed on December 6, 1995 was the 33,391-st full moon since the night of the Final Flyby.

Retro-calculating the position of the Moon to March 21, 701 B.C.E. agrees with the model to within .05 of a day, virtually to the hour. It also agrees with the ancient Jewish calendar to .05 of a day. Were this agreement be mere chance, be mere "coincidence", it would be one chance in 30. Such a coincidence is entirely possible. Whether or not it is "probable" depends on the evidence quality in the other eleven clues.

Clue # 2 - Dating The Final Fling Of Mars And Venus

This model for 8th century B.C.E. planetary catastrophism holds that the Final Fling between Mars and Venus preceded by about 56 days, plus or minus one, the Final Flyby between Mars and the Earth. The Mars-Venus flyby expanded the orbit of Mars, and thereby triggered the outside geometry of the Final Waltz of the Earth with Mars. If the first Mars-Venus flyby was the trigger, then the Mars-Earth flyby was the shot, the explosive event that produced gradualism for Venus, Mars and our planet.

Therefore this date, January 24, +/- 1 day, becomes one of the truly pivotal dates in the history of this Solar System. The retro-calculating of the position of Venus on January 24, 701 B.C.E. is as follows.

THE STARTING POINT. On December 8, 1874 there was an occultation, known as a "transit of Venus." Venus happened to pass directly between the Sun and the Earth. On that day both were at the same celestial longitude and latitude that particular day. The Earth, Venus and the Sun were in a straight sight line.

The Earth was at longitude 78 on December 8, 1874. Therefore, Venus in its orbit also was at longitude 78. This date and place (78° longitude) locates a "starting point" for retro-calculating the position of Venus back to January 24, 701 B.C.E.

The objective is to retro-calculate and locate Venus in its modern orbit, very early 701 B.C.E. Part one in this retro-calculation is to locate the position of Venus in late 700 B.C.E. By calculation the orbit in December 1874 was the 4,190th Venetian orbit since the day Venus seemingly acquired its modern orbit.

By the model, Venus changed from an old orbit period of 226.017716 days down to 224.701 days. The following is a retro-calculation of the position of Venus back to the 700 B.C.E.

For the preceding 4,190 orbits of Venus, from its 1874 transit, counting in Earth days, the multiplication is 4,190 x 224.701. The result is 941,497.19 days. In Earth years, this is 2,577.63 years. .63 years is also 230.1 Earth days. Following is the summary.

1. Dec. 8, 1874 to Dec. 8, 1 AD 1873 years

2. Dec. 8, 1 AD to Dec. 8, 1 B.C.E. is
(There is no year "zero" in this counting system.) 1 year

3. Dec. 8, 1 B.C.E. to Dec. 8, 700 B.C.E. 699 years

4. The Dionysius Exiguus factor 4 years

5. Dec. 8, 700 B.C.E. to April 22, 700 B.C.E. 230 days

6. Total 2,577 years 230 days

A second factor for consideration is transitional time. The "old calendar" contained 360 days, comprised of twelve months of 30 days each. (February 29 and 30 were real days to be counted in the old calendar and era.)

From this point, there is a retro-calculation of the position of Venus, two full orbits and 2°, which puts Venus at 76°. It is 722°.

A. The time Venus needed for 2 orbits and 2°:226.018 days = 452.036 days. 2° of additional travel for Venus is 2/365.256 x 224.701 = 1.23 days. 452.036 days + 1.23 days = 453.266 days

B. If 224.701 days, the modern period of Venus is used, the calculations are:224.701 x 2 = 449.402 days. Add 1.23days for Venus to advance 2°.The total is 450.63 days

C. The average between these two values for the transitional orbit of Venus is 451.948 days

----

Forward calculating for Venus is as follows:

a. January 24 to March 21, 701 B.C.E.
(counting 28 days for February) 56 days

b. March 21 to April 21, 701 B.C.E. 31 days

c. April 21, 701 B.C.E. to April 21, 700 B.C.E 365 days

d. Total - also 452 days

Thus, it retro-calculates that Venus was at 76° on January 24, 701 B.C.E., plus or minus 1 day. While the date is not known with precision, it is likely that the last celestial polka between Mars and Venus occurred on January 24, 701 B.C.E. January 23 and January 25 are possible.

The velocity of Mars as it went into its polka with Venus was approximately 97,100 mph. (Its velocity at its old perihelion nearby was slightly more, at 99,710 mph).

A COMPARISON - THE LAST POLKA AND THE FINAL WALTZ. Figure 23 (chapter 9) illustrates the geometry of the Final Fling of Mars with Venus, an inside flyby for Mars. Figure 24 illustrates the geometry of the Final Waltz of the Mars with the Earth, an outside flyby.

ENERGY. This model affirms that the Final Fling between Mars and Venus was only 64.82% as energetic as was the Final Waltz between Mars and the Earth. It was -.086479 energy units versus +.133424 units.

ANGULAR MOMENTUM. Similarly, there was 40.63% as much angular momentum exchanged during the Martian Final Fling with Venus compared to its Last Waltz with the Earth. It was -.008514 vs. +.020954 momentum units.

DISTANCE. The distance of the last Mars-Earth Waltz is estimated at 27,000 miles, center to planet center. The distance of the Final Mars-Venus Fling is estimated at 29,300 miles. The Final Fling was only 8.5% more distant.

Figure 25 illustrates the old orbit of Mars versus the new orbit of Mars. The old orbit of Mars has the longer "x" axis, a greater length. The new orbit of Mars has a longer "y" axis, a greater width. The old orbit of Mars had an eccentricity of .561 and swept out about 6% more space. The new orbit of Mars rounded out to an eccentricity of .093. The old orbit of Mars went out to the asteroid belt; the new orbit of Mars has ceased bothering either the Earth or Venus.

After the last Mars-Venus fling, 56 days later, plus or minus one, Mars assaulted the Earth, energetically but on the wrong side to maintain resonance. It was uniquely on the outside, farther from the Sun. The velocity of the red planet, as it began its last waltz with the Earth, was approximately 80,000 mph. The Earth's velocity was approximately 66,000 mph. Thus Mars lapped the Earth in space once again, but it would be the last time.

As was just mentioned, the span of time for Mars between the last polka, approximately January 24, to the last waltz with the Earth, late on March 20, was 56 days, plus or minus 1. Put another way, the Mars-Earth waltz occurred between 1,340 hours and 1,350 hours later than the Mars-Venus polka. The year 701 B.C.E. was a popular year for celestial dances, and this year was the finale.

CAUSE I. One cause for Mars lapping the Earth on the wrong side was that it gained a significant energy from its inside flyby of Venus. Mars acquired enough additional energy to expand its orbit. Therefore it lapped the Earth on the outside instead of the inside. Thus the Mars-Venus Fling was the trigger for the scenario; it (on January 24, 701 B.C.E.) was the beginning of the end for the Catastrophic Era.

CAUSE II. The second cause was the process of unraveling resonance. Computer analysis indicates that any outside flyby of the Earth will threaten to disrupt resonance. An energetic flyby will destroy resonance. Resonance was unraveled when Mars lapped the Earth on the "wrong" side. Mars had to make an inside (Sunward side) flyby for resonance to continue and the catastrophic era to be perpetuated. This time Mars didn't. Its pattern changed, bringing the end of an era.

CAUSE III. Another major factor in rounding out the orbit of Mars was Jupiter. Formerly, Jupiter (Zeus) played the role of holding Mars in resonance. Now out of resonance, Jupiter's gravity was a bigger factor than the Earth's gravity in forcing the orbit of Mars to circularize, to round out. See the Rudolphine Tables, XI, XII and especially Table XIII.

It should be added that this produced both positive and negative effects for life on the surface of the Earth. On the Earth's surface, short run, the catastrophic damage ceased, and this is the good news.

The bad news is that the dynamo of the Earth's geomagnetic field (Mars flybys) went dead. The geomagnetic sheath protects life on the Earth's surface from the actinic (short wave) radiation from the Sun. The geomagnetic field now has a half life of 1,350 years. In 701 B.C.E., its strength was 1.2 Gauss. Presently its strength is down to .3 Gauss, and will become .075 Gauss by 4700 AD Thus the days for life as we know it on the Earth's surface are numbered. Scientists need to understand this.

This study points out that a retro-calculation of the position of the planet Venus has been done. It indicates that in January of 701 B.C.E., on or about January 24, at the 76th longitude, the two orbits of Venus and Mars crossed. On this date, Venus was there and Mars also was there. Mars was on the inside; their closest distance was about 29,300 miles. Mars, the faster, lapped Venus from the inside, the sunward side.

Figure 24 - The Catastrophic and the Modern Orbits of Mars

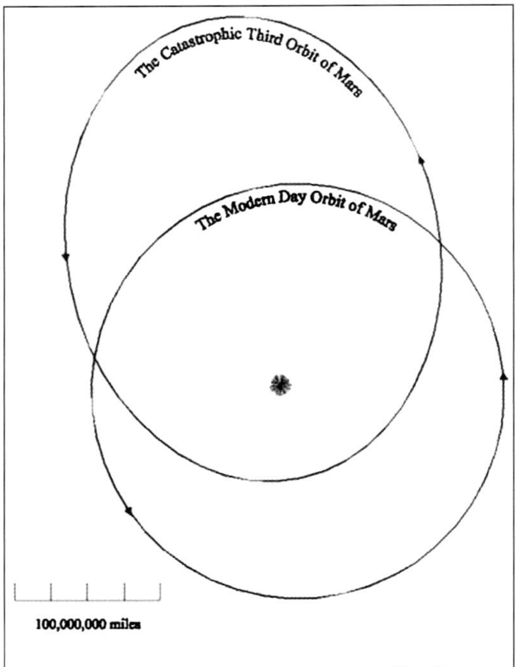

The Catastrophic and the Modern Orbits of Mars

Figure 25 - Geometry and Distances of the Final Mars Earth Flyby

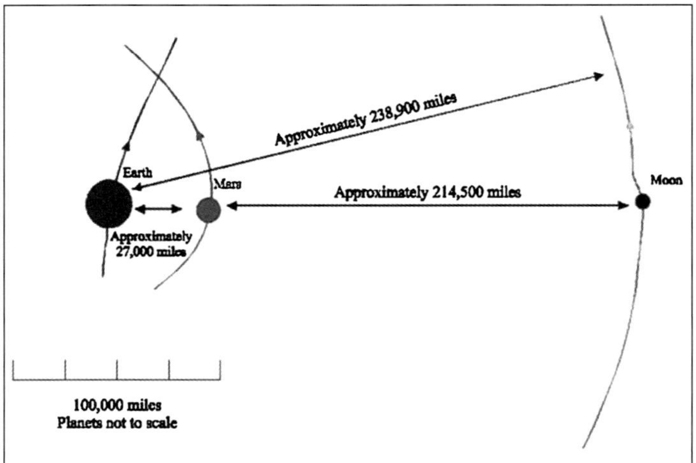

Geometry and Distances of the Final Mars Earth Flyby

First, this retro-calculation of the position of Venus agrees to within 1 day with the model of the Venus location.

Venus advances about 1.6° in one day. This intersection is the 76th celestial longitude, where Venus was on January 24, 701 B.C.E. Allowing for a margin for error of one day, this produces a 3.2-degree zone in the orbit of Venus. A retro-calculation of the modern orbit of Venus back to January 24, 701 B.C.E. agrees closely with the model.

Second, the chance of a retro-calculation of the Venusian modern orbit agreeing within one day with the model "by chance" is 360 divided by 3.2 or conservatively, one chance in 100.

In clue # 1, the chance of agreement by retro-calculation of the Moon's position was one chance in 30. The corporate chance that both the Moon and Venus agreeing is one chance, or coincidence, in 3,000. The chance for "coincidence" becomes somewhat weaker.

It is clear that this model of the old orbits of Mars, Venus and the Earth in Catastrophic Era CAN BE VALID ONLY IF THE TRANSITION MECHANISM WORKS IN ALL OF ITS PARTS. So far, the parts are beginning to "add up."

Clue # 3 - The Location Of The Perihelion Of Venus

The third interesting solar system condition to be noted involves the orbit of Venus, and its semi-major axis. The modern perihelion of Venus is located at the 131st celestial longitude. Its aphelion is at 311°.

The 131st longitude in the orbit of Venus coincides with the place in space where the old, catastrophic Mars exited from Venusian orbit space. Mars exited from the Earth's orbit space at the 180th longitude, which is our vernal equinox.

It is suggested that the very first Mars-Venus catastrophe, whenever it took place, was at longitude 131. It reorganized the semi-major axis of the orbit of Venus, an orbit which is close to circular.

By way of a parallel thought, it also is probable that the very first of the Mars-Earth flybys, whenever it occurred, was a March case flyby, on March 20-21, at longitude 179 or 180. That is where Mars put its first torque on the Earth's spin axis.

Ever since, the spin axis has been tilted so that its anniversary, March 20-21 has been the Earth's vernal equinox. March 20-21 was then and still is the first day of spring. It is one of the two days in the year when there is exactly twelve hours of sunlight in both hemispheres. (The other day is September 21, the first day of autumn). Thus it is suspected that Mars in its catastrophic orbit organized both the Venusian perihelion and the Earth's vernal equinox, which also was the Hebrew Passover and the Roman tubulustrium.

It appears now that the mechanism for the placement of the perihelion of Venus, at 131°, ALSO IS A RELIC OF THE ANCIENT MARS-VENUS WARS. This perihelion location is just as much a relic of the Mars-Venus Wars as are the scars on the surface of Venus, which also were Mars-induced.

It would never occur to a gradualist in astronomy that Mars orchestrated the perihelion of Venus. A gradualist has no reason to ask the key question. On the other hand, to a planetary catastrophist studying this model and the orbit of Venus, the orchestration of the Venusian perihelion by the catastrophic Mars verges on the obvious.

How do the traditionalists and gradualists explain this geometry of the Venusian orbit. The question has never surfaced until this time, until recognition of Clue 3. Explaining perihelion locations have not been a concern of gradualism. In the literature of astronomy, nowhere in the last 200 years has there been any explanation for the location of the perihelion of Venus, pointing to the 131st longitude. Until now.

Similarly, in all the literature of astronomy, there still (in 1996) IS no explanation that the Roche Limit of Mars caused the fragmented of Astra, creating the asteroids - and creating the Clobbered Hemisphere of Mars. Leaders in astronomy, highly intelligent to be sure, still are not asking the right questions.

The perihelion of Venus could be positioned at the place in space where Mars exited Venusian space, just "by chance," within 2° of either one of the two locations where Mars crossed the Venusian orbit. The chance of the Venusian perihelion aligning by coincidence with either one of the two ancient Venus-Mars intersection sites is a pair of four-degree chances. It is eight chances in 360, or one chance in 45. The corporate odds for all three clues being "coincidence" rises from one chance in 3,000 to one chance in 135,000, conservatively calculating.

Hesiod Reporting On Planet Wars In 701 B.C.E.

It is apparent from "The Shield of Herakles" that Hesiod saw the last polka of Venus and Mars (to him, a celestial war between Pallas Athene and Ares). He wrote about their celestial tiff, so he probably saw it, even though, at Thebes Greece, he was some 40,000,000 miles distant.

It would be the red planet's last tryst in space with Pallas Athene. However, as Hesiod saw and reported correctly, it was a celestial war between a fast, fired up, furious, impassioned Mars with a "scowling, gray-eyed" Pallas Athene. (Greeks had gender for their planets, masculine for Mars, feminine for Venus. He also portrayed the cloudy Venus as "gray-eyed," interestingly and in a sense, accurately).

Apparently, a powerful electrical flux tube formed between the two planets, possibly of over 100,000 volts. Lightning strikes swept between their surfaces for 10 or 15 hours. If so, it was similar to the "current" current in the flux tube actively and constantly flowing between Jupiter and its closest satellite, Io. Io is 255,000 miles from Jupiter, center to center.

The strength of the ongoing flux tube for electricity between Io and Jupiter has been estimated. If those estimates are correct, it flows constantly at 400,000 volts, 5 million amps, two trillion amps. The flux tube between Ares and Pallas Athene may have been less intense; it may have been more intense.

When lightning strikes, characteristically there is produced a crater with a small "hump" or rise in the center, at the strike site. Reportedly, there are many such "humps" in craters on the surface of Venus. If there was a flux tube between Venus and Mars, it would have lasted only 10 to 15 hours, when the two planets were in close proximity. If so, at 100,000 volts or more, it would have lit up both the surface of Mars and its cometary tail of Mars like a Christmas tree with 1,000 bulbs.

Today, an ongoing 400,000-volt flux tube flows across 255,000 miles of space between Io, the innermost Jovian satellite, and Jupiter. It is across a vacuum and normally lightning does not like vacuums. Jupiter is 390 times bigger than Venus, and it spins very rapidly, in less than 10 hours. Io may create vast amounts of friction within the rapidly-rotating Jupiter.

But on the other hand, Mars was eight times more massive than is Io, and on its final fling, at 29,300 miles, it was also momentarily eight times closer to Venus than is Io to Jupiter.

Thus our estimate of a series of 100,000 volt electrical discharges between Ares and Pallas Athene could be conservative. The Io-Jupiter flux tube was first photographed by Pioneer 10 and 11, 1972 and 1973.

If such a Venus-Mars flux tube suddenly appeared, it was due to sudden sub-crustal friction produced by each planet's gravity disturbing the fluid molecules in the internal regions of the other planet.

Clearly, Hesiod described an "aegis of Venus" when Pallas Athene got in the way of Mars. Unclearly, no translator of Hesiod understands what an "aegis" was, much less a flux tube, and much less the cometary tail of Mars, the "Fleece of Aries." Modern translators have no modern word for "aegis" so they use this Greek word without knowing its definition, or the background relating thereto. Thus it is that the "aegis of Venus", like the craters of former lightning discharges, is another of the many scars of Mars.

In ancient Greek literature, the cometary tail of Aries was well-known. In color, the

Fleece of Aries was somewhat silvery in color, somewhat golden, and always spectacular. On the light of these developments, it is known that Venus today has a thick gray atmosphere. Interesting it is that Hesiod described Pallas Athene as "gray-eyed." Once again, whether by chance or not, Hesiod as an ancient reporter, deserved a Pulitzer prize.

Perhaps this same "apparition," the cometary, wavy tail of Mars also was seen in the land of Israel. Mars had ices effervescing off only one hemisphere, its Eastern Hemisphere. Its ices effervesced in twelve-hour cycles. To the Hebrews, its cometary tail was seen and described as the fluttering wings of the destructive, dreaded Angel of the Lord. Wings of birds flap in cycles as apparently did the cometary tail of Mars.

Cometary tails today typically feature a twin symmetry in their tails, or "wings." Usually, with short term icy comet tails, the right side flutters in lock step and in mirror image with the left side. So it was that the fluttering, cometary tail of Ares resembled angel's wings just as much as it resembled the fleece of Aries (Ares).

> Listen to the words of our Greek eye witness:
> But now Athene, daughter of Zeus of the aegis came to stand in the path of Ares,
> Herself wearing the gloomy aegis.
> She looked scowling terribly at him and spoke with winged words,
> "Ares, stay now your fury and power, and your hands invincible; ...
> So she spoke, but could not persuade the great heart in Ares, but he, screaming aloud, flourishing his spear like a flame, rapidly made his rush against the powerful Herakles, furious to kill him ...
> But gray-eyed Athene, reaching out of the chariot, turned aside the shock of the spearhead. [n1]

The most important city in Attica then was Athens, a city named after Pallas Athene. In time, it became the capital city of Greece, and still is. Greek cosmic mythology casts Pallas Athene in the role of a "protectress" of Hera, the Earth. Hera and Athene were sister planets, both having endured assaults by Ares.

The conflict between Athene and Ares in January, 701 B.C.E., turned violent. Ares hurled fierce lightning bolts at Athene, and being bigger than Ares, apparently Athene hurled them right back. Again, Hesiod seems to have had it right.

As it turns out in Hesiod's "The Shield of Herakles," the first scene has Ares attacking gray-eyed Pallas Athene. But it was only the first act of a two act cosmic play ... a two act celestial drama. The second act was fireworks between Hera and Ares. According to the translator of Hesiod, Lattimore, Hesiod wrote at the turn of the 8th century B.C. and this model agrees. Isaiah lived then also.

Hesiod mentions the steeds of Ares in one way or another some 20 times in 480 lines of description. Phobos is less than 20 miles in its longest diameter and Deimos is less than 10 miles. Hesiod had their number right, two. He had their timing right, accompanying Mars during a Mars flyby. He had their color right, black. He had their speed right, a rapid rush (see line 452, The Shield of Herakles).

Hesiod's description is remarkable, but it would be even more remarkable IF Mars has not been nearby, in flyby phase. Gradualist astronomers are leaders for at least 98 of the cosmology taught today. Perhaps 1.9% aren't so sure. The leaders of ex nihilo creationism account for perhaps 0.1%.

Usually gradualists and ex nihilo creationists disagree on everything, especially when the dawn of Earth history occurred - 10,000 or 4.6 billion years ago. Regardless of when that was, those leaders are united in that, since the dawn of history, whenever it occurred, Mars has never been closer than 30,000,000 miles either to our planet or to Venus. Furthermore Mars has never been that close to the asteroid belt either. So the question is whether or not Hesiod saw the celestial scenes that he described.

Clue # 4 - RETRO-CALCULATING THE POSITION OF MARS BY MODEL THEORY Vs. BY OBSERVATION

In retro-calculating the orbit of Venus, the orbit of Venus shifted sunward only 314,000 miles. Venus has the velocity to make this shift in a few hours or days. Mars, on the other hand, shifted inward to an orbit averaging 141,600,000 miles but it last buzzed the Earth, at 92,250,000 miles.

The question is "Where was Mars on March 20, 701 B.C.E."? The retro-calculation of the position of Mars to March 20, 701 B.C.E., is as follows. Michelsen's Heliocentric Ephemeris places Mars at 10° Virgo on March 20, 1995. [n2] This position cited by Michelsen is also 160° longitude.

1. March 21, 701 B.C.E. to March 20, 1 B.C.E. - 700 years

2. March 21, 1 B.C.E. to March 20, 1 AD - 1 year

3. March 21, 1 AD to March 20, 1995 AD - 1994 years

4. Add 4 years for the medieval monk - 4 years

Total 2,699.000 orbits or years

5. In 2,699 Earth years, at 365.256 days per year, there are 985,826 days.

6. The modern period of Mars is 686.979 days.

7. Dividing 985,826 days by the modern period of Mars, 686.979, there have been 1,435.018 Mars years in 2,699 Earth years.

8. In retro-calculating the orbit of Mars, assuming no other factors were present, it back calculates to 154°. The model has Mars at 180° on March 21, 701 B.C.E. Mars is slow by 26°.

9. Mars advances .524° per day (360/686.978). 26°/.524 equals 49+ days. If this model is correct, 49 days is the amount time Mars required to find its new orbit. 37 or 38 days was taken by Mars getting to its modern orbit. This leaves 12 or 13 days to be charged to other causes.

10. At an average speed of 55,000 mph, to travel 49,300,000 miles, Mars would require 896 hours, or 37+5 days just to traverse the distance to its new orbit, if it proceeded directly. Such is unlikely, and a longer route is more likely, requiring a few more days.

A direct path accounts for 37 of the 48 days that Mars is short in actuality versus by the model. The rest, another 11 days, may have been consumed in "touring," approaching its new orbit but not by the shortest distance.

Mars advances .524° per day. To be cautious, perhaps was 11 days slow from our calculations. 11 days is 6°. For the model to be within 6° by retro-calculation, it is (360/6) one chance in 60.

Were this model without merit, Mars could just happen to be in the appropriate zone, Virgo, in March 1995, had it been orbiting in its modern orbit for billions of years, and were gradualism a fact of science. It would be one chance in 60 that the model just happens to retro-calculate well.

Corporately, including cases 1 through 3, it was found that the probability of coincidence was 1:30 x 1:100 x 1:45, or one chance in 135,000. Add the retro-calculation of Mars, one chance in 60. For the first four clues, or conditions to align this well with the model, 30 x 100 x 45 x 60, it is one chance in 8,100,000. Chance is fading as an alibi.

Clue # 5 - Jupiter's Orbit And Old Earth's Orbit

In this Solar System, many of the moons of Jupiter and Saturn are in various orbital resonances. 2:1 is the most common resonance. Europa and Ganymede are an example, as are Ganymede and Callisto. Mimas-Tethys, Enceladus-Dione are also examples. The period of the orbit of Mercury is in 3:2 resonance with its own spin rate. Pluto is in 3:2 orbital resonance with Neptune.

A few of the asteroids still are in resonance with Jupiter. The Trojan asteroids, for instance, are in 1:1 resonance. Three asteroids are in 2:1 resonance, China, Clematis and Griqua by name. The semi-major axes of these three asteroids line up parallel on a head to toe basis with Jupiter's semi-major axis.

There is one asteroid in 3:1 resonance with Jupiter, Alinda by name. Alinda has an average orbit period of 1,444.2 days; Jupiter's period is 4,332.6 days, exactly three times longer than Alinda's period average. The relationship of the semi-major axes of Alinda and Jupiter are of special interest. They are perpetually perpendicular. 3:1 resonance aligns this way. Apparently 6:1 resonance also aligns perpendicularly. Mars had 6:1 resonance with Jupiter.

It is apparent that in the Catastrophic Era, the semi-major axes of the Earth and Mars were aligned parallel, just like the 2:1 asteroids with Jupiter. What is not apparent, but is true, is that both the axes of Mars and the Earth together, in the Catastrophic Era, aligned perpendicularly with Jupiter, just like Alinda. There is evidence.

THE MODERN ORBITS OF THE EARTH AND JUPITER. Could it be that these two major axes of their orbits were in a perpendicular relationship? What is the geometry of their two semi-major axes in the modern age? The Earth's modern orbit is 365.256365 days. Jupiter's orbit is 4332.59 days. Today this ratio is 11.86 so they are no longer in resonance.

But if the Earth were to have had an ancient orbit of 361.1 modern days, they would have been in 12:1 resonance, with Mars at 722.2 days simultaneously in 6:1 resonance with Jupiter. At 365.256 days for the Earth's orbit, they no longer are in resonance. At 361.1 new days or 360 old days, they were, at 12:1.

Perhaps it seems complicated to the general reader, but it is all really very simple. In this era, the major axis of Jupiter's orbit is 13.6°, and it was the same in the Catastrophic Era. The Earth's orbit in the modern age is 102.2°. The difference between Jupiter's alignment and the Earth's today is 88.6°. This is only 1.4° from a perfect 90°, perfect right angles. The semi-major axes of the Earth and Jupiter still are virtually perpendicular.

If one asks where the perihelion of the Earth's orbit was in the Catastrophic Era, the answer is simple. It was half way between the October case flyby, October 24, and the March case flyby, March 20. This was 146 days, half of which is 73 days. 73 days from either flyby date, assuming all months had 30 days, was late January 5 or early January 6.

Figure 26 - Perpendicular Orbits - Jupiter and the Earth

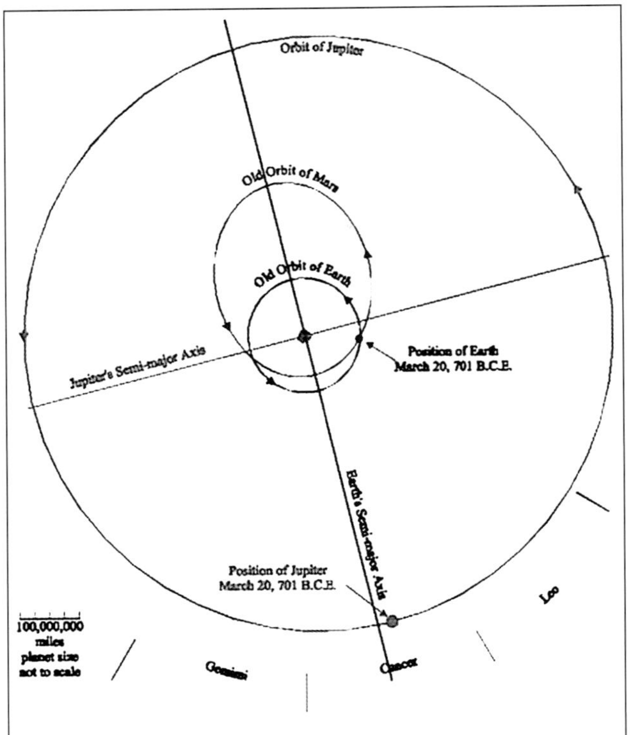

Perpendicular Orbits - Jupiter and the Earth

But today, some months have 31 days and one has 28. On today's calendar, the old perihelion would have been midday of January . In our era, it is January 3. This is another indication that when Mars yanked the Earth outward about 616,000 miles, it also yanked the Earth backward. Had it not been, the semi-major axes of the Earth and Jupiter today would be in perfect right angles, as they were in the Catastrophic Era. Figure 26 illustrates.

Gradualism dogmatists will allege that Jupiter's semi-major axis is nearly perpendicular with the Earth, by coincidence. Planetary catastrophists know better. It is just one more vestige, or relic of the Catastrophic Era. It is in the same category as the orientation of the Venusian perihelion to the exit site where Mars crossed its orbit.

What is the chance that by coincidence, Jupiter's and the Earth's semi-major axis align today within 1.4° of right angles? Conservatively, it is six chances in 360, or 1 in 64.

The corporate chance that the first four clues were all coincidences (30 x 100 x 45 x 50) was about one chance in 8,100,000. Multiply by 64. For all five of these clues to be coincidence, such would be the case once in 518,400,000, or approximately 500,000,000 times. "Chance" is becoming a poor if not an impossible explanation.

Clue # 6 Retro-Calculating The Position Of Jupiter During The Final Flyby

On the night of March 20-21, 1955 AD, Jupiter entered into the 5th degree of Gemini, which is longitude 245. Jupiter's period is 4334.649423 years. Calculations for the locating place in space on the flyby night for Jupiter on March 21, 701 B.C.E. are as follows.

1. March 20-21, 701 B.C.E. to March 20-21, 1995 was 2,699 years.

2. 2,699 modern Earth years = 985,827 days.

3. On March 20-21, 1995 Jupiter was at 245° longitude.

4. The year (orbit) of Jupiter is 4334.65905 days, per Table XIII. (The Internet fact sheet gives the orbit of Jupiter as 4334.319763 days).

5. 985,827 days / 4334.653907 = 227.4292 orbits.

6. .4292 orbit x 360° = 154.5°.

7. 245° - 154.5° = 90 or 91°.

The conclusion is that on the night of the Final Flyby, Mars and the Earth were waltzing at 180°, just entering the Final Flyby. At the same time, Jupiter was entering the zone of Capricorn, 90° behind the Earth's position, at right angles to the Earth and Mars, in a year with a catastrophe on schedule.

This means that Jupiter was in the position of being precisely perpendicular to the Earth (and Mars) during the Final Flyby. 108 is divisible by 12. It will be demonstrated in Volume III that March catastrophes were periodic; that period was 108 years to the day, nearly to the hour. Since 108 is divisible by 12, then the position of Jupiter was precisely perpendicular to the Earth and Mars during each and every March case flyby. This touch of geometry blends beautifully with resonance orbit theory.

Ancient Greek cosmology had Zeus, or Jupiter, as the choreographer of the cosmos. No doubt this is a result of earlier Chaldean cosmology, to which the Greeks fell heir. The Greeks, and the Chaldeans had it right. Modern gradualists have been oblivious to it all.

Resonance theory predicts that Jupiter, or "Zeus-pater" should be perpendicular in position, at right angles to the Mars-Earth flybys, including the last waltz. This retro-calculation of the position of Jupiter agrees with resonance theory. Jupiter was where resonance theory predicts it should have been.

Retro-calculation of orbits back to the catastrophic year 701 B.C.E. has been completed. The Moon has been back tracked and its position agrees. The position for Venus has also, as have been positions for Mars and Jupiter. They all backtrack well, very well.

Clue # 7 - Asteroid Orbits And The Hecuba Gap

There is another interesting clue of the former arrangement of the orbit of Mars. This clue is in the asteroid belt, the spray of asteroids out where Mars and Astra once roamed. When Astra fragmented on the red planet's Roche Limit, some 65% to 70% of the fragments missed Mars, and also avoided orbiting around Mars. They began to orbit the Sun. Such was the genesis of the asteroid belt.

As of 1996, more than 5,000 asteroids have been detected. Of them, the orbits for 1,000 are known. It is the distribution of those asteroids to which attention now is turned. Perturbations, largely from Jupiter but some low level ones, also perhaps from Saturn and Uranus, have rearranged the asteroids. Once they were in a spray distribution, but it is no longer so. Today they are in zones, with clusters and gaps. Figure 27 illustrates.

Note in Figure 27 that there are clusters of asteroids at certain resonances with Jupiter. The Trojan asteroids are two clumps that have settled in at 300 arc seconds, in 1:1 resonance with Jupiter. Another cluster is at 3:2. This is the same resonance as Neptune-Pluto.

However, of more significance are the gaps in the distribution of the asteroids. Jupiter, Saturn, Uranus and Neptune have created gaps at certain resonances. Four of the most prominent gaps in asteroid distribution are found at the following resonances: (a) 2:1 at 600 arc seconds, (b) 7:3 at 700 arc seconds, (c) 5:2 at 750 arc seconds and (d) 3:1 at 900 arc seconds. Collectively, the gaps in the asteroid distribution are known as the Kirkwood Gaps.

Attention is now directed to the 2:1 gap. Asteroids have piled up in a cluster on either side of this gap. Only three asteroids remain in 2:1 resonance, the aforementioned China, Clematis and Griqua. Of the two clusters, the inner cluster has piled up at a maximum at 570 arc seconds and the outer cluster's maximum is at 630 arc seconds.

In both cases, the new clusters peak at being out of resonance by 5%. This is where the new equilibrium occurs. It is difficult for an asteroid to leave resonance by just 1% or 2% or even 3% because Jupiter pulls it back into resonance. When asteroids go out of resonance with Jupiter at the 2:1 gap, they go out of resonance by 4%, 5% or 6%.This is their new stability is zone.

Now, attention is turned to the old orbit of Mars, its catastrophic orbit, and its period. It was in 6:1 resonance with Jupiter and simultaneously was in 2:1 resonance with the Earth, but due to the Final Flyby, its intensity and its unique geometry, resonance unraveled and

Mars found a new orbit. Mars found its new orbit period at 94.98% of its ancient orbit (686.978 vs. 723.257). See Table XIII, page 3, column 2.

What does this reflect? It reflects that when Mars left resonance with the Earth, and with Jupiter, it assumed a new stable orbit with a period just like the cluster of asteroids at 570 arc seconds. It is short of resonance by 5.02 % of its period.

By this example and Figure 27, Mars has a new orbit. Whether or not it went out of resonance with Jupiter, it certainly appears like it did. The 2:1 gap in the asteroids is only one gap; other gaps occur at 7:3 5:2 and 3:1. As is indicated in Figure 27, the name for these gaps is "The Kirkwood Gaps."

The asteroid gaps have various names. The name given to the 2:1 gap is the Hecuba Gap. In Greek mythology, Hecuba was the wife of King Priam, and the mother of such noted children as Hector, Paris, Helen, etc.

To make a long story short, Mars found its new orbit at a period just like the asteroids that also have left 2:1 resonance with Jupiter. Compared with Jupiter's orbit, the red planet's orbit now is 15.85% of Jupiter's orbit. The ancient 6:1 resonance period was 16.67%. 16.667 divided by 15.85 is at 6:1 resonance. 16.667 divided by l5.85% is 105.17%. Mars is now, 5% out of resonance with both Jupiter at 6:1 and with the Earth at 2:1.

To make a quick conclusion, if the orbit of Mars did not go out of resonance with the Earth's orbit, and with Jupiter's orbit, its orbit certainly looks as if it did.

Judging by the paucity of asteroids in the 2:1 resonance, and the numerous asteroids at about 5% one either side, it is judged that the chance that Mars never shifted into a new orbit from a former orbit, since the dawn of history, is estimated at one chance in ten.

At the end of clue # 6, the corporate chance for coincidence was calculated at one possibility in 500,000,000. Multiplying by 10, the corporate chance for all seven clues cited thus far is one possibility in five billion. This is one occasion in 5,000,000,000 for "gradualistic chance" to be a reasonable explanation. The standard, traditional menu is not good science. The number 5,000,000,000 is chances, not years.

Figure 27 - Asteroid Distribution and the Hecuba Gap of the Kirkwood Gaps

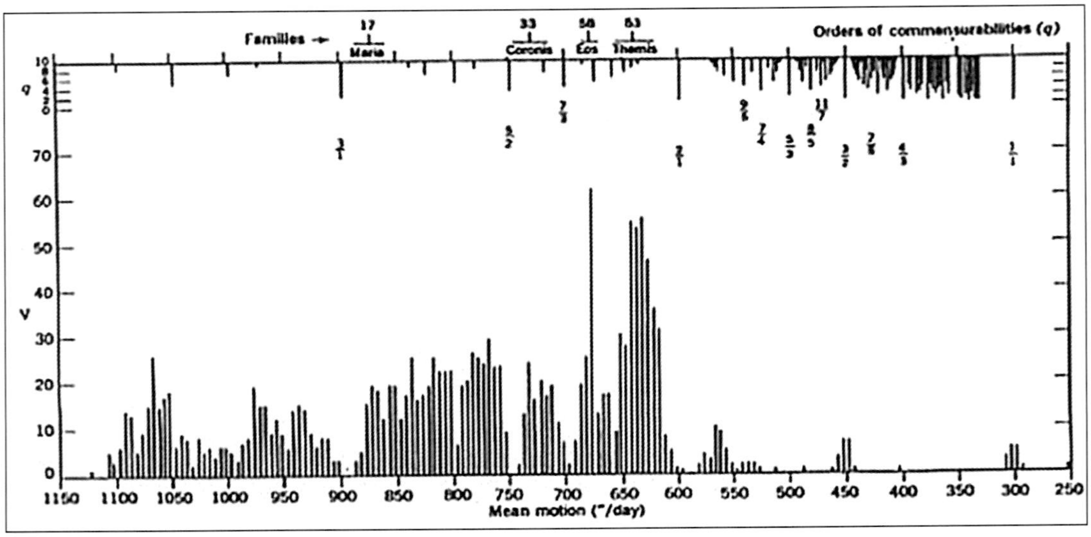

Clue # 8 - The Twin Tilts Of Mars And The Earth

So far, attention has been directed to various orbits such as the orbit of the Moon, the Earth, Venus, Mars, Jupiter and the asteroids. Changing now, clues 8 and 9 are directed toward the Earth's spin axis and the spin axis of Mars.

As was demonstrated in The Recent Organization of the Solar System, the spin rates are very similar, 1,436 minutes to 1,477. As was demonstrated in chapter 8, the tilt of these two spin axes also are similar. The tilts are 23.44° to 23.98°. The similarity of spin rate is 97.2% (spin axis rate) and 97.7% (spin axis tilt).

In Chapter 8, the conclusion was made that Mars-Earth torques occurred ALTERNATELY AND IN THE OPPOSITE SECTORS OF THEIR RESPECTIVE ORBITS. The two locations were March 20-21 and October 27. The longitudes were 180 and 33°. Indication are that the first Mars torque on the Earth was at the March 20-21 orbital location, and all subsequent March case flybys were odd-numbered. The second Mars-Earth flyby was at the October 24 site where Mars also crossed the Earth's orbit. All even numbered torques, beginning with two, were October cases.

Theoretically a 22.5-degree spin axis tilt is the ideal compromise angle under this scenario; it is half way between 0°, the most restful angle, and 45°, the most chaotic angle. At 22.5° is the most restful compromise angle. Both Mars and the Earth experienced energetic yanks of the final flyby. Even so, both spin axis tilts are close to that ideal compromise angle. This explanation for the twin angles of their tilts is the only explanation in the literature of astronomy.

How could these two phenomena occur by chance? It is estimated that for each planet, there is one chance in 20 of their being by chance close to this ideal compromise angle. In the light of the Mars-Earth Wars, the chance for these two tilts being twinned is astronomical; conservatively it is 20 squared, or one in 400.

The corporate chance for all eight clues being "coincidence" is 30 x 100 x 45 x 60 x 64 x 10 x 400. This is approximately one chance in 2 trillion.

Clue # 9 - General Spin Axis Precession And The Fixed Stars

In popular mystery stories, the more incidental and inconspicuous the clue, the better the mystery and the surprise ending. So it has been with Sherlock Holmes mysteries and Perry Mason performances.

So it is with the twin tilts in Clue # 8. Reciprocal, alternating torques between Mars and the Earth is the first and only explanation ever given in the history of astronomy for this twin phenomena. So it was with Clue # 3, etc. New thought and a new paradigm is needed for astronomy and cosmology, which is the history of the Solar System.

And so it is with the precession of the Earth's spin axis, clue # 9. In astronomy, the fixed stars are the most remote of phenomena, making this is a remote clue. In the past few centuries, the spin axis has pointed to Polaris, the North Star (Ursae Minoris). This star is 400 light years distant and puts out 1,700 times as much light as the Sun. Did the spin axis point to Polaris in 1,000 AD? No. In the time of Isaiah, Hesiod and Sennacherib? No.

Luni-Solar Precession And Planetary Precession

Precession of the spin axis addresses when and where the axis points in its circle among the fixed stars. The spin axis makes a grand circle requiring 25,800 years. It advances 1° every 71.667 years, about 1° in a life time.

Our best ancient sources indicate the ancient Greeks and Chaldeans had only 44 constellations, mostly in the Northern Hemisphere.

Today, and for the last 2,000 years, the north end of the Earth's spin axis points in general to Polaris, the North Star. Has the spin axis of the North Pole always pointed to Polaris? No.

Did it point to Polaris 2,700 years ago, in the age of Isaiah, Sennacherib and Hesiod? No. Then, the spin axis pointed to Beta Ursae Minoris, or Kochab. Kochab in Arabic means "Pole Star." Where will it point 7,000 years from now? Cepheus. Why?

The Moon's gravity pulls on the Earth, but not equally. The attraction is stronger on the equatorial bulge region, because the Earth is not a perfect sphere and there is about 1% more mass in the equatorial bulge zone. The Earth's equatorial diameter is 26 miles greater than is its polar (and spin axis) diameter. Thus the Moon puts a slightly stronger torque on the equatorial bulge. The Sun does also.

Because torques shift at perpendicular angles to the direction of the torque, the Earth's spin axis shifts 90° away from the direction of the Moon's torque. This force is called "lunar precession."

The Sun also pulls unequally on the Earth's equatorial bulge. Because of its greater distance, the Sun's force of attraction is slightly less than half of that generated by the Moon, but always, both forces work together. Their joint effect on the Earth's spin axis is termed "luni-solar precession."

However, there is also a third kind of precession, seldom considered as it is so tiny; it is "planetary precession". Planetary precession is caused by the remote planets, Jupiter, Saturn et al. This force is so minute it usually can be ignored. It is ignored because it is a minute effect in the present age, and it is assumed that what occurs in the present age also occurred in the past age. Wrong. In the past age, planetary precession was caused during the l08-year period of Mars flybys. It was sudden and was far greater than was luni-solar precession.

Due to luni-solar precession, the spin axis precesses slowly eastward. It requires 71.67 years, about an average life time, for the spin axis to precess one degree. In 25,800 years, the spin axis precesses a full circle in the heavens, 360°. Polaris has been the North Star for some 2,000 years. In 27,000 AD, Polaris once again will become the North Star.

However, overlooked by the astronomical community, during the Mars-Earth Wars era, planetary precession, caused by Mars flybys, overwhelmed all luni-solar precession. Mars planetary precession continually reset the spin axis back to where Kochab was. So Kochab continued to be the Pole Star for thousands of years, for the entire Catastrophic Era.

Through it all, one relatively tiny dim star of 4.2 magnitude remained as always being the first star on the eastern horizon at the moment of dawn of March 21, the vernal equinox.

The Dilemma Of Mesartim

The name of that small star is Mesartim. In ancient times, Mars-Earth Wars came and went. Latitudes could shift. Climates could change. Killer quakes could recur. Oceanic tides could overwhelm continental shores and sweep far inland. The crust could upthrust. Volcanoes could erupt and re-erupt during the next Mars flyby. Once an ice dump came, dumping celestially cold ice over our planet's two magnetic polar regions.

Cities or farmlands could be struck with celestial lightning. The cardinal directions could shift. The shadows on sun dials could relocate. But Mesartim continued faithful as the first star on the horizon at dawn, on the vernal equinox. It and the other fixed stars were about the only thing in the cosmos that didn't shift, shake, swell or spark.

Reference is made to Mesartim's small constellation. It is a modest, four-star constellation named "Aries." Aries is not large as constellations go, and its stars are not as numerous as stars are in other constellations. Its stars are not especially bright. But, like the month of January on the calendar, the constellation Aries it is well-placed.

This constellation is both the first and the leading constellation in the zone of Aries. Aries gives its name to an entire 30-degree sector of the fixed stars, the first slice of the zodiac. So, the zone of Aries, like January on the calendar, is the first zone in the zodiac. The constellation Aries is like the first week of January. The star Mesartim was well placed, like New Year's Day, the first star of the new year.

In the constellation of Aries there are only four stars, including Sheratan, magnitude 2.7, Hamal, magnitude 2.0 and Mesartim, a dim star, magnitude 4.2. Sheratan is the Arabic word for "sign". Hamal is the Arabic word for "sheep". Mesartim, the dimmest of the four, in Arabic means an especially fat ram. Mesartim, like January 1st on the calendar, is the first point of Aries.

Hamal, the Arabic word for "sheep, is the brightest in this small constellation, and has a magnitude of 2.0. It is sufficiently bright and prominent that navigators on ocean-going ships and submarines do use Hamal for determining location in the middle of the ocean at night.

Mesartim is both a dim star and a double star. In modern times, astronomers have discovered that it is actually a double star, one star virtually overlapping on top of the other. Together, their magnitude is only 4.2. Mesartim is not notable for its brightness. It is notable for its location. It is the first star in the 4-star constellation, Aries. Aries is the lead constellation in the 30-degree zone of Aries. The zone of Aries is the in the lead of twelve zones of the zodiac.

What is even more interesting and significant is that Mesartim ALWAYS WAS the "First Point of Aries," UNTIL AFTER 701 B.C.E. That was when Mars flybys ceased, and that was when planetary precession of the spin axis also ceased.

Aries is the leading constellation in the first zone of the zodiac. THUS MESARTIM

HAS ALWAYS BEEN KNOWN AS THE FIRST POINT OF ARIES since the dawn of Sumer and its zodiac. The zodiac, their map of the heavens, is the earliest idea from ancient pre-flood Sumer of which the modern age is aware.

In 1991, in our serene age of gradualism, Mesartim was listed in Bowditch's Book of Navigation Charts as being at 328°.F4 Hamal is at 332°. But in ancient times, Mesartim was always at 1° and Hamal at 5°. The positions of Hamal and Mesartim have shifted due to luni-solar precession over the last 2,700 years. They have shifted from the leading edge of Aries through Pisces and into Aquarius.

Mesartim was so important to the ancients because it always was the star on the horizon at dawn on March 21, the vernal equinox, the first day of spring and a new year.

Astronomers fail to recognize ancient Mars-induced planetary spin axis precession. Repeatedly, it overwhelmed the slow, ongoing luni-solar precession. Ancient literature, which is to the contrary of gradualism, is ignored.

Ancient sun dials, sun caves, sun spirals and other ancient devices for determining changes in latitude and changes in the cardinal directions, are not understood, and hence are usually ignored. Relocating the cardinal directions is what the remodeling of Stonehenge was about. Opinions are that Stonehenge was remodeled either four or five times. This suggests four or five spin axis shifts, or combinations thereof, that the Celtic astronomers considered major.

This also is what the obelisk construction industry in ancient Egypt was all about. It was important in that age to be able to measure and chart the shadow of the obelisk on key days like the summer and winter solstices, and the fall and spring equinoxes.

Mars shifted the position of the spin axis repeatedly, every century. In between, the first star in the zodiac (Mesartim) experienced a modest amount of luni-solar precession but then was reset by the next Mars catastrophe. Mesartim defined the beginning of the zodiac as much as January 1 defines the beginning of the calendar.

As it is with a calendar, so it is with every map of the world or of the heavens. There has to be an arbitrary starting place. For the modern calendar the arbitrary starting date is January 1 for whatever reason.

For the rotating Earth and its longitudes, the arbitrary starting place is the Greenwich Meridian, going through East London and the Greenwich Observatory. Greenwich is at zero longitude; there were laid out 180° of longitude on the Earth both to the east and to the west. They meet at the International Date Line. The total thus is 360°, just like the degrees in any circle.

So it is with the ancient zodiac, it also needed an arbitrary starting point. Mesartim filled that need very well. One reason the Greenwich meridian is located as it is at London is due to the location of an important 17th century astronomical observatory, the Greenwich observatory.

Another reason is the defeat of the Spanish Armada, which allowed England to become the queen of the seas. London became the foremost city of the world in banking, in overseas trade, in coinage, in colonizing, in slave trading, in map making and in map marketing.

For a brief time French cartographers placed Paris at 0°, but their maps didn't sell so well. German map makers centered 0° on Frankfurt. Portuguese map makers had Lisbon at 0° longitude. Dutch map makers used Amsterdam. But standardization was needed, and by 1750, the British system had gained world acceptance.

So it was with the map of the heavens. The zodiac, a map of the fixed stars, is the earliest idea that is known to have come from Sumer. It also occurs in India and China. The time selected to start the zodiac was the moment of dawn on March 21. This was "Mars' month" and the 30-degree zone of prominence was "Aries."

The First Point Of Aries

The ancient calendars had 360 days in Sumer, in India and in China. [n3] They needed surveys, and they all developed a 360-degree circle; the 360-degree circle also was applied to the heavens. It was subdivided into twelve 30-degree sectors, just like the months in some of the ancient calendars.

The word "zodiac" may or may not have originated in Sumer, but it clearly has come into English from Rome and Greece. In Greek, "zoa" means "the animals." Of the twelve sectors in the zodiac, eight are named after animals and four are named after kinds of people, as follows:

l. Aries, the ram	2. Taurus. the bull
3. Gemini, the twins	4. Cancer, the crab
5. Leo, the lion	6. Virgo, the virgin
7. Libra, the weigher	8. Scorpio, the insect
9. Sagittarius, the steed	10. Capricorn, the goat
11. Aquarius, the water bearer	12. Pisces, the fish

Mesartim Today

Today, in this celestially serene age, more attention is paid to the brighter Hamal than to the dim Mesartim. As was mentioned, Hamal is used by ship navigators at night and by nuclear submarine navigators also at night to locate their positions in latitude and longitude.

Published for over a century for the navy, Bowditch's Book of Navigation Charts, publishes star positions for oceanic navigation. As was mentioned, recently in 1991, Hamal is now listed at 332° (plus or minus a half degree) and Mesartim at 328°. [n4]

In the ancient times, Hamal was not at 332°; it was at 4°. Mesartim was at one degree, the first point in Aries celestially.

But this is an era where only luni-solar precession is calculated. For 2,699 years, both Hamal and Mesartim have been shifting slowly eastward, 1° every 71.67 years. In the 2695 years between 701 B.C.E. and Bowditch's 1991 publication, Mesartim and Hamal have slipped forward somewhere between 35(and 37°.

It is not known how much the spin axis was yanked backward by Mars in 701 B.C.E. as the Earth was also yanked outward. However a 2-degree yank backward is suspected based on the difference in the locations of the old and the new perihelions. [n5]

Astrologers follow a different zodiac, one that conforms to modern positions, and the history of that issue is not addressed here. The old zodiac is under discussion here. 2,699 years have produced 37.6° of shift. That is why Mesartim and Hamal have shifted, on the modern zodiac, out from Aries, all the way across Pisces, and into Aquarius.

Mesartim now is located at about 325.5°. It has shifted about 37°. As was mentioned, it shifts 1° in 71.67 years (25,800/360 = 71.67). Its annual rate of shift is 0.01395° (1/71.67). Luni-solar precession, uninterrupted by Mars planetary precession, is now 2,699 old, and counting.

The Dilemma In Dating Ancient Sumer

An interesting problem arises for anthropologists, archaeologists and some ancient historians who try to date Sumer. They tend to conform to the style of large year counts. Some like to date early pre-flood Sumer at 6,100 B.C.E. But if that dating were correct, and were luni-solar precession always continuing, in 6,100 B.C.E. then Mesartim would have been in Taurus. WHY THEN WAS MESARTIM CALLED "THE FIRST POINT OF ARIES"? Why would it not be called "the midpoint of Taurus"?

Other archaeologists date early Sumer at 8,500 B.C.E. Same question. Why did the ancients say that Mesartim was "the first point of Aries"? It would have been in mid Gemini.

Others, even more subject to gradualist dating, like to date Sumer when Mesartim should have been in Cancer, 11,000 B.C.E. Archaeologists who date pre-flood Sumer so early are sincere; they are merely conforming, and perhaps leading others in this style of dating.

It is the old question as to whether the emperor is or isn't wearing any clothes. Children and non-conformists say "No" with a giggle or a laugh or two. Archaeologists, like good citizens of the emperor's domain, maintain dates for Sumer, and Mesartim, that are curious. Why would Sumerian astronomers name Mesartim for a location it would eventually get to in 2,000, or even in 4,000 years into the future?

Archaeologists do not dare quarrel or challenge the principle of ongoing luni-solar precession; physicists and astronomers would have a fit. Nor do we. The only problem is that they overlook Mars planetary precession and the Mars-Earth Wars. There is quicksand For those who skirt this issue.

To date Sumer, one must also date the zodiac, one of the earliest of the Sumerian concepts. This includes locating in the cosmos the "First Point of Aries", Mesartim. Luni-solar precession has no chance as the sole explanation, excluding planetary catastrophism. It does not have even one chance in a million. The fixed stars, silent and remote as they are, also have their story for the court.

The first eight clues in combination gave "coincidence" one chance in 2 trillion. Multiply by at least a million. Coincidence and 18th-19th century gradualism now is calculated as having one chance in 2 quintillion of being correct. Conservatively, that is.

Clues # 10, 11 and 12 are developed in Chapter 12. Not all of the evidence is yet in court.

Conclusion

There is no scarcity of mathematical clues in this Solar System supporting this model of Mars-Earth Wars, and its extension, Mars-Venus Wars. Chapters 1 through 8 presented a series of clues from the physical geography of the badly battered Mars and its two severely pitted, poxed satellites. Gaspra looks like it formerly was with them.

Tables XI, XII and XIII present a logic of how both energy exchanges and angular momentum exchanges agree, and agree simultaneously, at every step.

Story 38 is that the Moon's orbit retro-calculates rather well with this model.

Story 39 is that the orbit of Venus retro-calculates very well also. Longitude 76 is where the Last Fling between Mars and Venus occurred, and January 24, 701 B.C.E. is the date, plus or minus one day.

Story 40 is that longitude 131 was the last exit location of Mars, crossing and leaving the orbit of Venus; it coincides with the Venusian perihelion. It is likely that the Venusian perihelion is explained by the first Mars-Venus flyby, with Mars exiting the Venusian orbit at 131°.

Story 41 is that Mars in its modern position also retro-calculates well with this model, featuring the Final Flyby, March 20-21, 701 B.C.E.

Story 42 is that Mars shifted out of resonance with both the Earth and Jupiter. This is the same ratio of shifts by the asteroids in the two Hecuba clusters of asteroids, surrounding the 2:1 gap. The new (modern) period for Mars shifted 5.02% in period away from the orbit of Jupiter, and 5.02% in period toward the Earth and the Sun.

Story 43 is that, during the Catastrophic Era, with 12:1 orbital resonance, the Earth's major axis was necessarily perpendicular to Jupiter's major axis. The Earth and Jupiter were in a 12:1 resonance. The Earth's semi-major axis still is within 1.4° of being perpendicular with the semi-major axis of Jupiter. This is another vestige of the Catastrophic Era, another relic in the cosmos.

Story 44 is that the Earth's spin axis has experienced only luni-solar precession in the modern era, over the last 2,699 years. But before that, planetary precession periodically overwhelmed and reset precession. In that era, dim Mesartim was repeatedly the "First Point of Aries".

By a study of (a) the chronicles of the Hebrew kings, Judah and Israel, the date of the Final Flyby can be ascertained, March 20-21, 701 B.C.E. By (b) retro-calculating the position of Jupiter, it can be shown that Jupiter was in Capricorn at 90°, on the date of the Final Flyby. This harmonizes happily with resonance theory.

By (c) a study of luni-solar precession, the end of the Catastrophic Era can be approximated, though it cannot be precisely dated. It coincided with the beginning of the modern era. Other resources are sufficient to tie down the Final Flyby occasion date to the fraction of a day.

The least familiar, and the most remote, but probably the most daunting of all of the evidence presented in this chapter is the modern position of Mesartim, now 37° from its ancient Sumerian mooring. At 71.67 years per degree there has been 2,700 years of luni-solar precession uncorrected by planetary precession. The luni-solar precession, uncorrected, is the timer.

The ancients had a lack of faith in the Sun rising from the same place on the horizon, and the stability of the cardinal directions over the centuries. This belief among people of the 20th century AD is taken for granted. In the modern era, this fits, but in the Catastrophic Era, it was not the common experience or expectation.

This understanding of the Mars-Earth Wars puts a new light on the extreme interest the ancients had in the constellations and in tracking the planets in their courses across the heavens. It puts a new light on their interest and practical use of their zodiac, their map of the heavens.

Story 45 is that with this map of the heavens and with accurate historical and astronomical records, prophets, monthly prognosticators, star gazers, astrologers, astronomers and swamis could accurately assess both (a) the date of the most recent celestial holocaust and (b) the timing of the next one. Studious prophets could predict when the next Mars holocaust would arrive to the day. That is why their advice was so valuable, and that is why they sat next to kings in the ancient royal councils.

With story 45, the reader now is 87% of the way to the penthouse of planetary catastrophism. Like it is at Seattle's famous Space Needle, the view is majestic, and the menu is good too.

CHAPTER 12

Clues Ten, Eleven And Twelve

> *Any idea that appears to challenge orthodoxy and is published is usually accompanied by copious responses from the names in the profession who are given an opportunity to quash any heretical conclusions which the article might suggest. Many subjects, no matter how interesting, are simply prohibited because they call into question long-standing beliefs.*
>
> Vine De Loria Jr. [n1]

Clue # 10 - The Ancient Preference For 360-Day Calendars

The tenth clue in this case is the ancient preference for 360-day calendars. It was not for 365 or 365.256-day calendars. And also there was a preference for 360-degree circles in the Middle East, the Near East and the Orient. After 701 B.C.E., only the Chinese changed their circle to 366°, the current count. They were, in a sense, more consistent than Western calendar makers and astronomers. But then, consistency has not been a necessity for mankind.

There are a minimum of fourteen ancient civilizations that used 360-day calendars in catastrophic times. Often they featured 30-day months, and twelve of them per year. For this to be accurate, the Earth's orbit would have to have been some 92,250,000 miles from the Sun, not the modern 92,959,000 miles.

There are more than fourteen ancient 360-day calendars; other researchers are welcome to expand this ancient day count per year. 360-day calendars have been found to have been useful in such ancient societies as Arabia, Assyria, Chaldea (Babylon), China, Egypt, Ethiopia, Greece, India, Israel, Japan, Mexico (the Mayans), Persia, Peru (the pre-Incas), and ancient Rome, for starters. [n2]

But gradualists assume, with excessive confidence, that the year has always contained 365.256 days since the dawn of time. They also assume that Mars has been only a small pinpoint of light in the heavens, never closer than 30,000,000 miles and never farther than away than 240,000,000 miles. This system of thought indicts ancient calendar makers as having not been able to count.

What is the chance that none of these ancient calendar makers could count to 360 correctly? Further, what is the chance that all of them couldn't count? Further yet, what is the chance that in their errors, they all made the same error? All gradualists and traditionalists in cosmology would have mankind to assume the following.

1. The ancient astronomers of China could not count. They had 24 fifteen-day periods based on a lunar calendar, from new moon to full moon, and from full moon to new moon. Neither did their ancient 360-day circle have any significance other than that it made the same error as did their calendar. The chance of this being true is estimated, conservatively at one in a hundred.

The English translations from the 11th century Soochow astronomical charts and the common Chinese names with English translations are given in the table below. The number of degrees in each tables (sic) does not add up to 360 because in the 4th century B.C. the Chinese astrologers suddenly changed the division of the circle from 360° to 365° 15 minutes. [n3]

2. The ancient Japanese calendar makers could not count any better than their Chinese counterparts. This chance again is remote.

3. The ancient calendar makers of Vedic India could not count days in the years, despite their very accurate system for counting the seconds in a day. Their smallest unit of time was .4 of a second, the vipalas. 60 vipalas, 24 seconds, was a pala. 60 palas made one ghatika, 24 minutes. 60 ghatikas made one day. 360 ghatikas originally made one year. Texts in the Veda know only of 360-day years. The chance ancient Brahmans could not count is also assessed at one in a hundred.

4. The ancient Persians could not count. In the Bundahis, an ancient sacred book, there were 180 days, not 182 or 183 days, from winter solstice to summer solstice. Later, after the 7th century B.C.E., they added five "gatha days." The chance that the ancient Persian calendar makers could not count is also assessed as remote.

5. 360 days made an old Babylonian year. What is the chance that ancient Babylonian/ Chaldean calendar makers could not count the days in the year correctly? Their mathematicians understood irrational numbers like Pi (3.1416) and Phi (1.618) and square roots. In Volume III, it will be clear that at least one of their engineers (Noah) knew about pi, phi (.618) and its square (.382). Sumerian mathematicians couldn't count? They counted with a sexigesimal system. The chance that they could not count is beyond remote.

6. The ancient Assyrian year also consisted of 360 days, and a sarus was 3,600 days, ten years. Their lunar month also was thirty days, crescent to crescent. The chance that their calendar makers could not count is estimated at one a thousand.

7. The ancient Mayan year was a "tun," 360 days. They had five-day periods, and 72 of them consisted of their year. They had better calendars than the Europeans up into the 19th century. What is the chance that Mayan calendar makers could not count? One in a thousand? Hardly.

> ... Sylvanus G. Morley and Herbert J. Spinden, both studied this matter extensively and found themselves in disagreement over the epoch, and hence over the Gregorian conversion of the calendar. ... They did agree that the epoch and the date when the calculations first became effective are two different things, and that some drastic alterations in the counting system occurred between the 6th and 8th centuries B.C., ... [n4]

8. The ancient year of the pre-Incas in Peru was 30 days to a quilla, and twelve quillas to a year. What is the chance their calendar makers could not count? It is illogical.

9. The earliest Romans reckoned 360 days as a year, according to Plutarch in his "Life of Numa." Other sources agree. Could not the Romans mathematicians count up to 360? The chance they couldn't count is one in thousands.

10. The ancient Egyptians originally had a 360-day year in their Book of Sothis. They could construct pyramids and could approximately calculate the diameter of the Earth. The innocent assumption that they could not count is ridiculous.

This 360-day calendar, like so many others, was changed during the 8th century B.C. to one of 365 days. The extra five days were simply added to the end of the year. [n5]

11. The ancient Greeks also had a 360-day year until the time of Thales (640? - 586 B.C.E.). Thales was among the first to recompute the calendar at 365 days in the late 7th century B.C.E. Why did the Greeks make the same error a dozen other ancient societies made? The chance of the Greeks being unable to count to 360 (or 365) : one in billions.

12. The ancient Arabs had a 360 day year, and moreover, they erected 360 idols to "Hobal," cognate with the Chaldean Bel and the Phoenician Baal (Ares-Mars). It was one idol for each day. Why did the ancient Arabians make the same error that calendar makers in a dozen other societies also made?

13. The year of the Hebrews, up until the 7th century B.C.E., consisted of 360 days and twelve months each of 30 days. During or after the 7th century B.C.E., a five-day period was added to Adar, their twelfth month. Veadar was 5 days long.

What actually happened is that the 7th century B.C.E. was a time when calendar makers all over the world were kept busy trying to develop a new calendar that would serve well for a century or longer. Adding five days per year was a stop gap; it meant the new calendar was 23 or 24 days in error after a century. Fair, but not good enough.

There are three issues embedded in this discussion. The first is (a) that ancient calendar makers, mathematicians and star gazers from at least different societies all could not count. Neanderthals or less. The second issue is (b) they all made the same mistake. The third issue (c) is that some of those societies employed 360-degree circles as well.

In clues one to nine, it was calculated that if chance explains them all, it is one chance in two quintillion. Add three more zeros for thirteen 360-degree calendars, and that is conservative. Next, add another three zeros for their all making the same mistake, and that is restrained and discreet. Finally add three more zeros for their 360-degree circles. And that is restrained and temperate. Now, "coincidence" calculates at one possibility in 2 octillion. It is one chance in 2,000,000,000,000,000,000,000,000,000. This calculation is discrete, restrained, moderate and reasonable.

Clue # 11 - The Missing Clue - Swift's Laputan Astronomers

In 1726, Jonathan Swift wrote the famous book, Gulliver's Travels. Swift was an Irishman, a Protestant who objected strenuously to the unfair treatment of the Catholic Irish by the English nobility in Ireland. Swift was also a satirist and a boisterous brother in London's John Bull Club. Swift enjoyed good jokes and satires, especially when they were about the hypocrisy of the proud, the powerful and the prosperous upper classes in the England of his day.

In Gulliver's Travels, there are at least three themes involving satires. In the person of Lemuel Gulliver, Swift was poking fun at English society. One is tales of the little people of Lilliput. Another is about the giant people of Brobignag. The third is about the astronomers of Laputa, an island off the coast of China. The astronomers on the pleasant island of Laputa in the early 18th century had superior telescopes to those inferior ones of the Europeans; thus they had a better knowledge of the cosmos.

For instance, the Laputan astronomers knew what astronomers of European didn't, namely, that little Mars had two tiny moons. (Swift was writing in London where among his friends were Isaac Newton, Edmund Halley and William Whiston, all renowned mathematicians, all competent astronomers and one or two were also members of the John Bull Club).

These two tiny moons were first viewed and discovered in the modern era in 1877 in a new telescope at the Naval Observatory in Washington D.C. by Asaph Hall. To the discoverer goes the privilege of naming them, a tradition in astronomy.

Accepting the advice of an Englishman, familiar with ancient Greek mythology, Hall gave the two new satellites of Mars the same names as did the ancient Greeks - Deimos and Phobos. This was a coincidence. In Greek, these words meant, and still mean "fear" and "panic," with different inflections of the verb, to fear, or to dread. In this way, the two unnamed moons of Mars which Hall discovered were given the same name as the ancient Greeks name for them, Deimos and Phobos, fear and panic.

Phobos, the inner one, irregular in shape, somewhat fragment-like, has three diameters, 12.5 miles x 14.25 x 17.5. The three diameters for Deimos, also fragment-shaped, are 6.25 miles x 7.5 x 10. These two tiny moons were well beyond the power of European telescopes and American telescopes until late in the 19th century.

Nevertheless, 151 years before Asaph Hall's discovery, Swift wrote about two tiny moons that Mars had, as follows:

> ... they have likewise discovered two lesser Stars of Satellites, which revolve about Mars, whereof the innermost is distant from the Centre of the Primary Planet exactly three of his Diameters, and the outermost five; the former revolves in the Space of ten hours, and the latter in twenty-one and a half; so that the squares of their periodical times are very near in the same proportion with the cubes of their distance from the center of Mars, which evidently shews them to be governed by the same law of gravitation, that influences the other heavenly bodies. [n6]

Swift's satire of 270 years ago has provoked a variety of comments by the gradualist establishment. All are statements of amazed disbelief. One of them is that if the Laputans had good enough telescopes to discover Deimos and Phobos, why hadn't they also discovered Uranus? Why hadn't they discovered Neptune? And more satellites of Saturn? Uranus wasn't discovered by Herschel until 1781, with a homemade telescope. Neptune was discovered in 1846.

Table XIV - Laputan Measurements Versus Modern Measurements - Orbits Of Phobos And Deimos

Satellite	By Swift's Laputans	By Modern Astronomers
Phobos	3 Mars diameters 12,420 miles period - 10 hours	.4 Mars diameters 5,760 miles 7 hours 39 min. [n7]
Deimos	5 Mars diameters 20,700 miles 21.5 hours	3.5 diameters 14,500 miles 30.3 hours

Table XIV gives the measurements of the orbits of Deimos and Phobos, according to Swift, as held by the Laputan astronomers, 150 years before their discovery. Table XIV also provides the modern measurements by astronomers of the 20th century.

Who in 1726, in England, with a knowledge of both Kepler's laws and Newton's laws on planetary motions, could have performed these calculations for Swift? Swift was a friend of Edmund Halley of comet fame, also of Isaac Newton of gravity fame, and of William Whiston, also a mathematician, astronomer, a friend of both and a catastrophist. Any one of this trio understood Keplerian celestial mechanics, the Newtonian laws, and could have assisted Swift in the calculations. Someone other than Swift had to couch the information in the language of mathematics.

RICHARDSON ON SWIFT'S SCENARIO. According to a mid 20th century astronomer, Swift said and implied the following:

> ... the squares of their periodical times are very near the same proportion with the cubes of their distances. ... There is no escape from the conclusion that Swift must have had to calculate the period of the outer moon from the equation:
> (Period of Moon) 2 = 53 / 3 X 102 = 12,500 / 27 = 462.97
>
> which gives: period of moon = sq. rt. 462.97 = 21.5 hours, very closely. Swift was a well-educated man, such elementary arithmetic should have been easy for him, although his principal training had been in literature and history.
>
> All this has been known and commented upon for nearly a century now. What seems to have been overlooked is that we can learn still more about Mars from the information Swift gave us about his moons. For when he specified the periods and distances of the moons he necessarily specified something else about Mars too. He specified its mass. [n8]

First of all, it is reasonable to disagree with Richardson on one point. This math was too difficult for Swift without help from someone knowledgeable in Keplerian and Newtonian thought.

At that time, in 1725 and 1726, astronomers did not know the diameter of Mars. Laputans disclosed the distance of Phobos and Deimos from Mars not in English miles but rather in Mars diameters. Astronomers in the early 1700's did not know the accurate value for the length of the astronomical unit, or how far the Earth is from the Sun. And they didn't know how far Mars was from the Sun. This unit of measurement in the satire suggests a very ancient sketch was involved, or a copy thereof from the Catastrophic Era.

This floats two questions. (1) Where did the sketch come from, and what was on it? (2)

Who, then, did Swift's calculations for him? Swift was an Irishman, a satirist and a wit. A mathematician he was not.

London at the time had a population of perhaps 200,000. That is where the John Bull Club met, and among its members were Swift, Isaac Newton and Edmund Halley. The probability is that Swift's mathematician/astronomer was either Edmund Halley (1656-1741), Isaac Newton (1642-1727) or less likely, William Whiston (1667-1752). Flamsteed is also a possibility.

ISAAC ASIMOV ON SWIFT'S SCENARIO. What is the explanation of Swift's satire? Is it really a great, flamboyant coincidence? Such is the conclusion (guess) of Isaac Asimov, prolific writer on scientific themes, recently deceased. He confidently writes:

> This is an amazing coincidence. Of course, Swift might have reasoned as follows: It was known that the Earth had 1 moon, Jupiter 4 and Saturn 7 at the time he was writing his book. It was reasonable to suppose that Saturn might have an 8th moon hidden somewhere and, in that case, if Mars had 2 moons, there would be a nice list of numbers.
>
> As one moved outward from the sun, beginning at Earth, the number of moons of each planet would be 1, 2, 4 and 8. Then, too, the moons of Mars would have been small and close to the planet, or even Europeans with their 'poor' telescopes would have discovered them.
>
> So far, Swift's thinking can be followed. However, his guess that Phobos would rise in the west and set in the east because of its speed of revolution is uncanny, IT IS UNDOUBTEDLY THE LUCKIEST GUESS IN LITERATURE. (Our caps.) [n9]

Asimov was the one doing the guessing ... guessing that Swift had been guessing. The following may be an inkling as to Swift's source. It is known that in the late 1610's, the works of both Kepler and Galileo, along with Copernicus, were banned by the Vatican for Roman Catholic scholars.

THIS TEAM'S PERSPECTIVE ON SWIFT'S SCENARIO. It is also known that by the mid 1600's, Jesuit scholars were paying no attention to the Vatican's ban. Jesuit missionaries were active in the Far East, and specifically in Peking, the Chinese capitol.

Despite the Vatican ban, from their school Jesuits adopted and taught the Keplerian approach, Vatican bans or otherwise. They were trying to collaborate and integrate Western science and Chinese science. Many including Emperors Shun-chih and K'ang Hsi, were considering converting to some form of Christianity. Among their reasoning was that Western astronomy (i.e. Keplerian thought) was so much better than traditional Chinese astronomy.

The names of some of those Jesuit missionaries include Frs. Matteo Ricci, Johann Schreck, Adam Schall von Bell, Wenceslaus Kirwitzer, Paul Hsu, Ferdinand Verbiest, Jean de Fontaney and Jean-Francois Foucquet. Any of these, but especially Verbiest and Foucquet are candidates as the one who might have made a copy of an ancient sketch of Mars, with Deimos and Phobos, to Paris, their home base.

It is possible that in the astronomical archives of Peking in the 1670's and into the 1710's, there were copies still in extant of ancient star charts of Mars with Phobos and Deimos circling it. If so, they were copies of ancient Chinese star charts over 2,400 years old at that time.

If such sketches existed, very possibly one of the Jesuit missionaries seeing such a sketch of the ancient "Fire Star" would be puzzled by Mars with two tiny satellites. Perhaps a copy of the copied original sketch arrived in Europe by the early 18th century. Swift's close friend, a Catholic, Alexander Pope, had Jesuit contacts.

Jesuit archives of the early China missionaries need scrutinizing for pertinent information. If such a sketch was there, the Jesuit fathers in Peking would have been puzzled, and the Jesuit fathers in Paris, from where Verbeist and Fontaney came, would not have known what to make of such a sketch.

But if a copy of such a sketch fell into the hands of Swift, Newton or Whiston, perhaps at the John Bull Club, some interesting interpretations could have resulted, and laughs also. A scenario like this, it is suspected, is where Swift got his "inspiration" about the Laputan astronomers and their understanding of Mars with satellites.

To Swift, at the very least, such a sketch could provide the basis for an excellent satire on the "backwardness" of English academia. They weren't as intelligent as they thought they were. They needed to have some fun poked at them.

Charles McDowell has researched this subject, and he suspects that the transmission of a Mars-Deimos-Phobos sketch, or sketches, came from Peking via a Dr. Cunningham to Dr. Hans Sloane. Sloane gathered samples of thousands of new plants from newly explored lands, for his greenhouse and his herbarium. Sloane's personal collection of 50,000 books was purchased by parliament in 1753, and became the core of the British Museum. [n10]

McDowell thinks "Lemuel Gulliver" was a cryptic name for Sloan, Queen Anne's physician. "Lemuel" in Proverbs was a wise man, one close to publishers and to the royal family.

Deimos And Phobos In Ancient Literature

CHINA. In his doublespeak satire, Swift virtually announced to any baffled cryptographers that his source was China; and China meant the astronomical archives of Peking. Indications and suspicions have been voiced, but the hard evidence is still missing. It may exist in Sloane's collection in the British Museum.

GREECE. Hesiod's treatise, "The Shield of Herakles" (Hercules) is only 480 lines long. Panic alone is mentioned twice (lines 144 and 236). "Panic and Terror" are cited once (line 463), as is "Terror and Panic" (line 195). Interestingly for revolving satellites, "Onrush and Backrush" is his translation once (Line 154).

"Fast-footed horses" (lines 95-97) is a translation used once, as is "horses on either side" (line 346, of Ares) once. "Feet of the horses" occurs once (line 63). "Running horses" occurs in Line 62. "Straining horses" occurs in line 65. "Fluttering-maned horses" occurs in line 372. [n11]

The inner steed, Phobos, has a period of 7.65 hours, and thus traverses 180° from one side of Mars to the other, a half orbit in only 3.8 hours. It rises on the western horizon of Mars, and was portrayed in that manner by Gulliver/Swift's story.

Deimos has an orbit in 1.26 days, also quite fast, and rises in the East, like most satellites. (Mars, like most planets, rotates from west to east). "Both horses" is translated once (line 346). This puts the count for the steeds pulling the chariot of Ares at two, no more and no less. The count for references to celestial stallions, either Deimos, Phobos or to both, is 20 to 23 times in 480 lines.

"Ares" is mentioned by name 13 times and archetypes of Mars are repeatedly mentioned, names like Gorgon, Typhon, "Shaker of the Earth," "War God," "Manslaughtering Ares," "Grim-faced Ares," Perseus, "Powerful son of Zeus," "Lord of Battles," and "Ares the Destructive." "Deathmist" is Lattimore's translation in line 264, and may refer to Mars with its cometary tail in that age, their "Fleece of Aries."

Zeus is mentioned repeatedly, as is Pallas Athene (including her gray eyes). "Apollo," a borrowed word from Phoenicia, for "Baal," is used once, and Aphrodite also appears once. In general, it is a description of planetary conflicts. Many think it was of one conflict between Ares and Hera, but this model suggests it was two conflicts. The first was between Pallas Athene and Ares; the second between Hera and Ares.

ISRAEL. Were the Bible organized chronologically, the first book listed would be the Book of Job. Some of its more intriguing discussions are on ancient cosmology:

Canst thou bring forth Mazzaroth in his season? or canst thou guide Arcturus WITH HIS SONS?

Knowest thou the ordinances of heaven? canst thou set the dominion thereof in the earth? Job 38:32-33

ARCTURUS. The English word translated "Arcturus" is from Latin, and probably is Greek in origin. In Hebrew, the word translated "Arcturus" is 'ayish or 'ash, which seems to be the old Hebrew word for the red planet, Mars. When the adjective 'mah (the great, the mighty) is added to 'ayish, it becomes 'mah 'ayish, "the great Mars," or "the mighty Mars."

MAZZAROTH. In Hebrew, "retet" and "retheth" are words for terror, trembling. "Mizrach" means "from the east," or from the direction of sunrise. Thus "mazza-retheth", or Mazzaroth, meant "the terror from the east," that is, from where the Sun rises, the celestial east in space. (Since the Earth rotates from west to east, all planets and stars rise in the east.)

"Arcturus," in Hebrew, 'ayish, is almost certainly a planet, not a constellation. Constellations do not have sons or satellites, and distant stars do not have binary partners that were visible to the naked eye.

The number of "sons" or satellites of 'ayish is not cited in the Book of Job, but "two" is an excellent guess. Elsewhere in an obscure Talmudic passage, the count of two is cited for the sons of Arcturus. The Hebrew word "aur" is cognate with the earlier Sanskrit word for Mars, "ayish".

If it is granted that Swift was privy to a copy of an ancient Chinese sketch of Mars and its two moons, there becomes three ancient places as sources for the two moons of Mars, not one (Greek sources). One source might be in a pictographic language, the "Fire Star," the Chinese source. The second is a Greek source; Greek is an Indo-European language that eventually was alphabetized. The third source is a source from the ancient Semitic family of languages, the 38th chapter of the Book of Job.

It is the suspicion of the authors that "Arcturus and his sons" as a translation was just an ancient translator's guess. A better rendition would be "Mars and its two satellites," or alternately, "Ares and its two steeds."

If and when the missing clue surfaces, a suspected sketch of Mars, it will make three ancient sources from three ancient languages for the orbiting Deimos and Phobos. It will also create "panic and terror" in gradualism's cosmology.

Earlier, the chance of "coincidence" was calculated at one possibility in two decillion. The chance that some form of an ancient Chinese sketch will turn up is much better. A search should be put on in archives of Peking, Paris, the Vatican, monasteries of Belgium and Germany and the British Museum, London, with attention given to the papers of Edmund Halley, Isaac Newton, Alexander Pope, Dr. Hans Sloane and Jonathan Swift.

If such a search is successful, Marsographers and planetary catastrophists should not be surprised.

Clue # 12 -Literary Scars Of Mars In The English Language

369 Words in English Derived from the Red Planet

(80% Relate to "Bad Action" or "Bad Experiences")

Individuals carry both physical and psychological scars. So do societies. Physical scars are the more obvious, and heal faster. It is the psychological scars that, often, are deeper and slower to heal.

So it has been with ancient scars on mankind from the ancient Mars-Earth Wars. Mankind has been afflicted with episode after episode of ancient catastrophism - all close flybys by Mars. The scars remain in our calendar (March). They remain in the theme of holidays (Halloween and Passover). The scars also remain words derived from the planet Mars in the modern English language. The large majority of those words can be characterized as "bad action" words.

Fifteen of the physical scars of Mars have been discussed, and include:

The fragmentation of Astra on the Roche Limit of Mars.
The evidence that Mars put on 1.5% more weight.
The resulting genesis of the asteroids.
The sudden genesis of the former Martian ring system.
The numerous pitlets on both Deimos and Phobos.
The numerous pitlets also on tiny Gaspra.
The crater distribution on Mars, 93% on one side.
The rim where the Martian hemisphere of craters ends.
The genesis of its two bulges, Tharsis and Elysium.
The birth of its giant rift system, Valles Marineris.
The birth and infancy of the giant volcanoes of Mars.
The dry river beds on Mars (but without any canals).
The cause of twin spin axis tilts - Mars and the Earth.
The genesis of the ancient icy cometary tail of Mars.

The psychological scars of the ancient Mars-Earth Wars also have caused very deep memories with long-lasting roots imprinted on our language. The following is a list of 369 words in modern English, which are derived from a variety of ancient words for Mars and its satellites, Phobos and Deimos.

Note that most of these words listed below usually connote some kind of a "bad action" or a connection with a bad experience. Many come from Latin. Many come from Greek via Latin and French. Some of the words have Germanic, Greek, Hebrew, Phoenician or Vedic origins.

Catastrophic Words In English Derived From The Words Ares, Astra, Bel, Indra, Mars, Tyr, Etc.

FROM GERMANIC NORTHERN EUROPE (Via North Italy). Tyr, the Scandinavian deity of battle = Mars. In Anglo-Saxon, Ti, Tiig, genitive, Tiwes, whence, 1. Tuesday - Mars' day, following the Sun's day and the Moon's day. In old Norse, Mars was tivar, Lithuanian, devas, old Latin, dius. 2. Deity, Deism. 3. Deify. 4. Trauma. 5. Trepidation. 6. Tempest. 7. Tremor. 8. Trouble. 9. Turmoil. 10. Turbulent, Turbulence. 11. Tumult. 12. Turbid. 13. Turbine. 14. Trouble. 15. Tumble. 16. Trundle. 17. Trumpet. 18. Tremble. 19. Torch. 20. Temblor. 21. Terror.

One of the two Roman days set aside for fearing and venerating Mars was on March 21. Its name was their "tubulustrium," an anniversary day remembering with horror some of the past cosmic holocausts. "Tubulustrium" in Latin was from Etruscan roots, and is cognate with the words listed above.

From Chaldea, Phoenicia And Carthage And On To Rome

During the youth of the Roman Republic, Roman traders exchanged merchandise, money, letters of credit, words and phrases with Carthaginian merchants. Carthaginian and Phoenician ships traveled great distances across the Mediterranean Sea, across the Indian Ocean and apparently to a limited extent, also across the Atlantic Ocean.

Then came Rome's two Punic Wars with Carthage, long, dragged out affairs, culminating in the demolition of Carthage. Carthage had been a colony of Phoenicia. Phoenician was Carthage's language. There was ample time and occasion for Phoenician loan words to be borrowed into Latin. The Phoenician word for Mars, "Baal," was one of those loan words.

BAAL = Mars. FROM CARTHAGE. In Latin, "bellum" was the word for war. Bellona was the Roman goddess of war. 22. Bellicose. 23. Belligerent, Belligerence. 24. Ante-bellum. 25. Hannibal and his father Hasdrubal, both Carthaginian generals during the Punic Wars; both fought Rome and both were named after Mars.

FROM PHOENICIA. 26. Jezebel and various Phoenician place names, such as 27. Baalbek, etc.

FROM CHALDEA, where BEL = MARS. 28. Belshazzar, 29. Babel.

From Greece

In Greece, "Apollo" was a Hellenized loan word from the Phoenician "Baal" (Mars.) It was Hellenized with the softening of the "b" to a "p" plus additions of a prefix of "a" and a suffix of "o". From Greek, 29. Apollo, as in the Apollo space program. 30. Apollos, an early Christian leader of prominence, and perhaps "apologetics."

FROM GREECE - TYPHON, via Persia and to China, tai fung, a great wind. 31. Typhoon. "Typhon" was a name for the type of cosmic storm, moving from east to west as the Earth rotated in a marine setting. It described Typhon's journey above and across the Mediterranean Sea during Mars flybys from the east (the Levant) to the west (Sicily.)

Apollodorus described Typhon vividly. Typhon was a son of Hera but was without a father. He was a grisly monster with 100 dragons heads, cast into Tartarus by Zeus. He was the father of the savage Cerberus, the horrible Hydra, and the catatonic Chimaera and of dangerous winds. Typhon is related both to the Chinese t'ai fung, a great wind, and the Arabic tufan, also a great, tempestuous wind.

FROM GREECE. ARES = Mars (from Sanskrit "ayush.") Attributes: a celestial warrior with a spear, many flashes of cosmic lightning and a burning torch. The hurling of his torch or celestial spear was the signal for the commencement of celestial warfare. Many of these words come to English directly from Latin, and others via Latin and French.

32. Aryan (Arian). 33. Ardor. 34. Arm. 35. Armada. 36. Armor. 37. Armistice. 38. Army. The second Roman fear day was October 24, their "Armilustrium." It is cognate with the above words. 39. Areopagite, originally court of astrologers, seers, star gazers and monthly prognosticators who met in Athens on Mars hill, near the Acropolis to prognosticate. Later, early in the modern era, the Areopagites became Athenian philosophers and judges.

40. Arrant (errant), a wandering vagrant of notorious character. 41. Arrogant, arrogance. 42. Arrow. 43. Arsenal. 44. Arson. 45. Arsenic. 46. Artillery. 47. Argonaut. 49. Armature. 50. Error, related to errant, arrant.

FROM DEIMOS, A SATELLITE OF MARS. From Deimos, the smaller, outer satellite. Deimos is a Greek word for terror. 51. Demon, demonic. 52. Demented, dementia.

FROM PHOBOS, the inner of the two ancient steeds of Ares. In Greek "phobos", like Deimos, was another of Greek words for various aspects of panic, fear and terror. 53. Phobia. 54 - 309. Roget's International Thesaurus lists 256 different kinds of phobias including kleptophobia, the fear of stealing, hydrophobia, the fear of water, Anglophobia, fear of the English and cometophobia, the fear of comets.

Mars was a comet as ices effervesced off the surface of its Eastern Hemisphere during the Catastrophic Era. With the rounding out of the orbit of Mars, nevertheless, a fear of comets continued, even into the medieval age. Cometophobia began to end with the coming of Kepler, Galileo, the telescope, Halley of comet fame and Newton.

One significant well-known, historical example of cometophobia was the appearance of Halley's Comet in 11 B.C., the Bethlehem Star. It coincided with the birth of Christ. The shepherds saw the star and succumbed to cometophobia.

Also consumed with cometophobia, the Magi journeyed from Babylon to Jerusalem, and Bethlehem, in search of perhaps a new king and a new dynasty, perhaps the promised Messiah. In Jerusalem, cleverly, King Herod asked the Magi to report back to him whatever information they found. (They didn't.)

The Roman puppet "king", Herod, enraged and afflicted with cometophobia, had all of the male babies under two years of age in and around Bethlehem slaughtered to ensure the death of the rumored child of the heavenly sign.

If the population of Bethlehem and environs was 10,000, and if the average length of life was 50, that would have been a slaughter of some 200 male babies. This royal rampage, too, was a result of the prevailing cometophobia that afflicted King Herod. Cometophobia was a psychological scar of Mars.

The next appearance of Halley's Comet was in December, 69 AD It was held to mean the coming of a new dynasty. As a result of a certain astrologer's advice, Nero slaughtered numerous potential rivals in Rome who he or the astrologer deemed to be a threat to the line of Caesar.

Comet appearances in Rome also were considered to herald the coming of a new dynasty, bad news for the old dynasty. Paul, an early Christian leader, already incarcerated in the Mamertine prison, was among those that Nero slaughtered. Nero and his court magician, like King Herod, were afflicted with cometophobia.

In Jerusalem in December of 69 AD, astrologer types also saw this appearance of Halley's Comet. Their interpretation also was that it was a sign of an upending of the Caesars and Roman rule. It was the ideal time to rebel from Rome; even the signs in the heavens were favorable.

With this interpretation, evident from the heavens, astrologers badly advised the leaders of the rebel party. It was a sign from heaven to the zealots of the upcoming fall of the Roman Empire. The zealots were the party of that era, in the name of patriotism, planning a violent revolution against Rome. They guessed, wrongly, that it was a good time to capitalize on the troubles of Rome.

Thus it was that the rebel zealot leaders were persuaded by "a sign from heaven" to publicly rebel against Rome; the populace either agreed or were intimidated. The upshot was a ruinous rebellion, at least as devastating and sad as was our Civil War. The astrologers and the zealot leaders, in 69 AD in Jerusalem, also became victims of cometophobia, and it was a sad turning point in Jewish history. In all three cases, in Bethlehem, in Rome and in Jerusalem, massacres resulted from a widespread cometophobia.

From Egypt

HORUS was the Egyptian Mars. It is cognate with the Greek Ares and the Hebrew mah 'awr. Greece also borrowed loan words from Egypt and vice versa. 311. Horror, horrid, horrible. 312. Hours and 313. Horoscopes.

The ancient Hebrew word for Mars seems to have been something like "'owr" or "'uwr", with the prefix "'mah." Thus "'mah 'owr." Thus it was that the Egyptian word and the Hebrew word for Mars were similar, which is not surprising, since early on, Israel spent many years in Egyptian servitude. Egyptian was the first language of Moses. These words were cognate with the Sanskrit "ayush," the Greek Ares, and the planet Mars.

From Rome

"Mars" apparently was a Latin slur of the Greek "Magna Ares," the great Mars. Mars was the nominative case, martis was the genitive form. 314. The month of March, one of the two months that featured Mars flybys. 315. To Mar. 316. Marauder. 317. Martial. 318. To March. 319. Murrain, a cosmic pestilence.

Mars flybys brought to the Italian peninsula widespread damage, death, terror and destruction. The Latin word and deity of death is "mors" (genitive mortis) and this word is related to (and cognate with) the genitive form of Mars (martis.) 320. Morose. 321. Mortal. 322. Immortal. 323. Morbid. 324. Moratorium. 325. Morgue. 326. Mortgage. 327. Mortify. 328. Mortician. 329. Murder. 330. Murmur. 331. Morsel. 332. Mourn.

The Latin word for a star, either a moving planet or a fixed star, was "astrum," (masculine) and "astra" (feminine.) The Latin "astra" was cognate with the Phoenician planetary deity, Astarte (Venus.) Apparently "Astarte-Astra" became Phoenician loan word to the Romans, like "Baal-Apollo" was to the Greeks.

333. Asterisk. 334. Asteroid. 335. Astonish. 336. Astound. 337. Astral. 338. Astrolabe. 339. Astrology. 340. Astronomy. 341. Star. 342. Astrophysics. 343. Catastrophe, catastrophist (a planet approaching, throwing down bolides and/or bolts of electricity from above). 344. Disaster. 345. Diastrophism. 346. Disarray.

From India

In Indic, INDRA = Mars. 347. India, the land of Indra, 1,600,000 sq. miles, 3% of the land area of the Earth. 348. Indians. The populace inhabiting old India, including citizens of Pakistan, Nepal, Bhutan and Bangladesh, exceeds one billion. This is 22% of the world's population.

349. The East Indies. 350. Indonesia. These geographical words also are cognate with Indra, the Vedic Mars, Indra. Indonesians number another 200 million, named after the Vedic Mars.

351. American Indians, the name given to the native race inhabiting the Western Hemisphere. This race was misnamed by Christopher Columbus, because to his dying day he thought he had discovered a route to the East Indies. Columbus' name stuck. American Indians comprise some 150 million, or 3% of the world's population. As a group, "Indians" from India, Indonesia and American Indians total over 25% of the world's population.

The Roman word for the planet "Venus," as it so happens, is derived from the Indic "Vishnu." Why the Romans adopted a loan word for one planet from India, and another loan word for another two planets from Greece is a good question. Vishnu probably was an Indic loan word to the Trojans, and from there to the Etruscans, before Rome was founded, circa 750 B.C.E. At any rate, there was no law saying that the Romans had to be consistent with their sources for loan words.

352. THE INDO-ARYAN RACE, Part I. The Brahmins of India are Indo-Aryans and they claim that their civilization is the oldest of all civilizations, - even older than Sumerian civilization. There is a line of logic indicating that this may be true; ancient Sumer may have indeed been a branch of a more widespread paleo-Indic civilization, or perhaps it was vice versa. In the West, Sumer has been considered the most ancient of the literate societies.

The "Indic" part of "Indo-Aryan" is from Vedic sources, Indra-Mars. The "Aryan" part of is from Greek and perhaps Hittite sources, Ares-Mars, which is related to the Sanskrit "ayush". Due to these circumstances, this hyphenated name, "the Indo-Aryan race" is the linguistic equivalent of "the Mars-Mars race." These words are also scars of Mars.

Add the India Indians, the Indonesian East Indians, the American Indians and the Indo-Europeans and the total population of these races is some 60% of the world population. Even a greater percentage of the world population speak one or more of the Indo-European family of languages. This is because English has become the world language of business and science.

353. The Indian Ocean. This ocean, named after the Vedic Mars, contains 25% of all the water on the surface of the planet Earth. By coincidence this proportion is similar to the

per cent of the world's population that are from the sub-continent of India and Indonesians.

354. The East Indies, a geographical term for a collection of islands.

355. The West Indies, another geographical term.

356. Indiana, a state and a county in Pennsylvania.

357. Indianapolis, the capitol of Indiana.

358. Indian River, a county on the Atlantic coast of Florida. In addition there are such towns as:

359. Indianola, in Washington, Nebraska and Mississippi
Indian Harbour, Labrador
Indian Head, Saskatchewan
Indian Hills, Colorado.
Indian Lake, Michigan.
Indian Springs, in California and Virginia
Indian Trail, North Carolina
Indian Village, New York

FROM NORTH CAROLINA. 360. Mars Hill, a village in North Carolina named after 361. Mars Hill, a hill in Athens prominent in the Book of Acts. This was where the Areopagites, seers and monthly prognosticators of Ares, once counseled citizens of the city of Venus (Athens) regarding past and future catastrophes and of their celestial schedules.

ANCIENT CHINESE. Ancient Chinese writing, a pictographic language, is understood today only with difficulty. The modern Chinese character for the planet Mars means "the Fire Star." The ancient character for the fire star meant the same thing.

According to our advisor on the Chinese language, the ancient Chinese word for Mars contains two characters, each character of which has a top part and a bottom part (like fractions). The bottom part of the first character seems to mean "intent" or "purpose." The top half he could not understand. Of the second character, the top half means "fire" and the bottom part again he could not understand.

Yet he confirmed to us what the Chinese language signifies, namely that Mars was indeed a fire star with some kind of a purpose, perhaps also a schedule. This ancient pictograph for Mars is thought to antedate 2000 B.C.E.

362. Leviathan, a large and dangerous dragon. This was the name of a great ancient monster of the celestial deep, which was Mars by another name. For people of the Near East, Leviathan came from the celestial deep, not from the Indian Ocean. Leviathan is

cognate with Lotan, an ancient Greek term for a cosmic dragon.

363. Behemoth (behemaut.) The terrible Behemoth was another horrible dragon of the great celestial deep that sometimes devastated the Near East region of the Earth, or even the entire planet's surface. It could be that Behemoth made its awful, occasional appearances and noises in the Hebrew month of Nisan, roughly March, while Leviathan made its even more devastating appearances in the other catastrophic month, October.

Consult the Book of Job, chapters 38 and 39. Teammate, Rehab and Asp were other Hebrew names for a dragon of the great celestial deep, and are probably from Chaldea.

FROM HEBREW. 364. Cherub. An angel, a messenger of the Lord of hosts. The following description of a "cherub," is hardly a lovable tot; it occurs both in Psalm 18 and II Samuel 22. The following is a citation from the King James Bible.

> Then the earth shook and trembled; the foundations also of the hills moved and were shaken, because he was wroth.
>
> There went up a smoke out of his nostrils, and fire out of his mouth devoured: coals were kindled by it.
>
> He bowed the heavens also, and came down; and darkness was under his feet.
>
> And he rode upon a CHERUB AND DID FLY, yea, he did fly upon the wings of the wind.
>
> He made darkness his secret place; his pavilion round about him were dart waters and thick clouds of the skies.
>
> At the brightness that was before him his thick clouds passed, hail stones and coals of fire.
>
> The Lord also thundered in the heavens, and the Highest gave his voice; hail stones and coals of fire.
>
> Yea, he sent out his arrows, and scattered them; and he shot out lightning, and discomfited them.
>
> Then the channels of waters were seen, and the foundations of the world were discovered at thy rebuke, O Lord, at the blast of the breath of thy nostrils. Psalm 18:7-15. (See II Samuel 22:8-16, a duplicate passage.)

In the widely regarded Strong's Concordance, an interesting comment is made.

3742. ..... keruwb, "ker-oob." of uncertain der. a cherub or an IMAGINARY figure.

Strong was mistaken; a keruwb was not an imaginary figure. Its crust was solid, rock solid; it had a mass 11% of the Earth' mass, and it had a cometary tail that extended tens of millions of miles. 365. Cherubim. 366. Sennacherib. (In Assyrian, "Sin" was the Moon, thus Sennacherib was named after the Moon plus the dragon Mars).

367. Iran, Iranian. Like Ares, Iran is cognate with, the Sanskrit word for Mars, which is "ayush", or "arush." As was mentioned previously. the English word Aryan (Arian) is derived from this Sanskrit word for Mars. Iran is just another form of "Aryan." In 1931 the ancient name of Persia was changed officially to Iran. This indicates correctly that the ancient peoples who settled Iran were primarily Aryans, not Semites. Aryan (Iran) at that ancient time was the land of devotees to, and followers of Ares.

The early post-flood settlers of northern Persia, for the most part, were the Medes, one of the earliest of the Indo-Aryan peoples. "Medes" and Medea are both cognate with Madai, a grandson of Utnapishtim (Noah.)

368. Arch, from Greek meaning "chief, great, principal." 369. Arc.

Conclusion

Until a sketch of an ancient Chinese star chart surfaces, or a copy thereof, it is premature to consider as hard evidence Swift's Laputan astronomers and their knowledge of Mars, its diameter and its two satellites. But there is ample circumstantial evidence, however, to suspect such a sketch existed and still might be found.

Story 46 is about the ancient 360-day year in over a dozen ancient calendars. The gradualists assume a 365+ day year has been installed since the beginning of time, and the calendar makers in this dozen plus of ancient societies all were mistaken. Moreover, they all made the same error. These ancient calendars are evidence of the ancient 360-day orbit of the Earth-Moon system.

Story 47 is that there are at least 369 words in English that are derived from various ancient words for the planet Mars. Some 10% of them are geographical place names. Well over 75% are "bad action" words, derived from catastrophic or cataclysmic themes. Enlil, Ares, Bel, Baal, Indra, Mars, Phobos, Typhon, Deimos, Horus, Cherub, Behemoth and Leviathan all have provided roots for words used in modern English.

There are concepts in the heavens such as the Greek "fleece of Aries," and the "aegis" of Venus. There is the "fluttering wings of the terrible angel of the Lord," the pestilence, fire, brimstone, the killer quakes, terrible thunder. To the ancients, such themes were scary, and they applied to scenarios, repeating century after century.

Definitely these words were "bad action words", from catastrophic experiences. The derivation of the English word "cherub," a toddler with a sweet disposition, is misleading. "Cherub" also came into English, via Hebrew, from catastrophic scenarios 3,000 and 4,000 years ago.

Sometimes, experts in academia make condescending remarks about how primitive were the ancients of the Catastrophic Era, and how fearful and superstitious they were. This implies those modern experts, knowing no cosmology other than gradualism, suppose they know more about ancient astronomical conditions than did the ancients themselves. 41. Arrogance.

If one chooses to doubt the menu - the conclusions and the dogmas of ancient gradualism, as your authors do, he or she then becomes skeptical of the traditional skeptics. But what is the word for that? Advocate of a paradigm shift? Believer? Catastrophic cosmologist? Iconoclast? Marsographer? New fashioned? Old fashioned? Planetary catastrophist? Student of ancient cosmology? Super skeptic? Perhaps as in mathematics and geomagnetism, the double negative becomes the positive.

With story 47, the reader is 90% of the way to the penthouse of catastrophic cosmology. There are only seven more flights of stairs to climb. These will be found in Volume III, "The Flood of Noah." It will be an analysis of the closest of all of the Mars flybys. The suspected distance, planet center to center, 15,000 miles! The suspected date, October 24, 2484 B.C.E., a day on which the Ark was ready, but just barely.

For the Mars-Earth relationship, with Mars less dense than the Earth, the Earth's Roche limit is about 10,500 miles, and perhaps as far out as 11,000 miles. If Mars came within 15,000 miles of the Earth's center, it was 4,000 miles from its Roche Limit, less than one Mars diameter. On this closest of occasions, unlike Astra, Mars survived, just barely. Also, the Sumerian Utnapishtim, who is the Hebrew Noah, and his wife also survived, also just barely. It was no pleasant weekend cruise.

Volume II And Its Place In The Series

Volume I addresses the evidence of a recent flare up, or nova of the Sun and recent delivery of the planets to the Sun, and the Sun's capture of them. "Recent" is used in the astronomical sense, not the historical sense.

Volume II, agreeing with Homer, indicts Mars as the "bane of mortals". It addresses both the evidence and the clues that Mars was indeed the "bane of mortals" on the Earth, the bane of Astra and the bane of diastrophism for the crust of Venus as well. It discusses how, and to some extent when, Mars and other planets were delivered to the Sun.

Chapter 7 addresses the Flood of Mars, and the genesis of the short term icy comets of the Solar System.

Chapter 8 addresses the twin tilt phenomenon between Mars and the Earth.

Chapters 9-10 address how the catastrophic era orbit of Mars rounded out, and was converted into its modern low eccentricity orbit. Also addressed was how the Moon shifted from a 30-day orbit to one of 29.53 days, and how the Earth shifted from a 360-day orbit to one of 365.256 days.

Chapters 11 and 12 provide an abundance of psychological clues, more than twelve, that as a package support the model and also indict gradualism.

Volume III, targeted for 1997 or 1998, is designed to analyze the closest ever of all of the Mars flybys. It was late in the year (October 24) of 2484 B.C.E. It addresses where Noah's Ark was built, and why. Also addressed are such topics as the ark's size, dimensions and shape, its construction site requirements, how far was it relocated geographically, in which direction according to modern coordinates, and where it finally was moored.

Are there geological relics in India that are parallel to the Ark's journey into Central Asia? It will address the route the floating Ark took, and how fast it moved.

Also to be addressed will be how deep the flood waters were, the peak velocities of those titanic tides and their source. Genesis cites that source as "the great fountains of the deep." The modern geographical word for that source is the Indian Ocean.

Volume III also will involve an analysis of the ancient periodic cyclicism of Mars flybys, 108 years, March case to March case, and 108 years, October case to October case, to the day and almost to the hour. The primary evidence is from the chronologies of the Old Testament, where chronologies are intermixed with accounts of catastrophic scenes. Material by Josephus and Talmudic material add to the scenario. In all ancient literature, only in the Old Testament and related Hebrew writings are catastrophes interleaved with chronicles.

Secondary evidence also will be cited from architectures, literatures and traditions from a dozen non-Hebrew sources from at least four continents. Those continents will be Africa, Asia, Europe and North America. Those societies will include the Mayans, the Vedic Indians, the ancient settlers of Cambodia, the Tibetans, Germanic sources (540 gates to Valhalla), Roman sources, pre-Inca sources perhaps, Japanese sources, etc.

Volume IV, if time allows, will address the Mars-Earth Wars from perspective of the Earth's physical geography, and the damage our planet has endured. The damage to the surface of Venus also may be a topic, as will the Earth's geomagnetic field, and the electrical aspects of Mars-Earth flybys.

Endnotes

Chapter 1 - Endnotes

Chapter 1, Note # 1 Alexander Heidel, *The Gilgamesh Epic and Old Testament Parallels*. 1946, Chicago, Univ. of Chicago Pr., p.87.

Chapter 1, Note # 2 Louis Ginzberg, *The Legends of the Jews*. 1909, Philadelphia, The Jewish Publishing. Society of America, Vol. I, p. 162.z

Chapter 1, Note # 3 *Loc. cit.*

Chapter 1, Note # 4 Ginzberg, *op. cit.* Vol. V, p. 164.

Chapter 1, Note # 5 Ginzberg, *op. cit.*, Vol. I, p. 105.

Chapter 1, Note # 6 Ginzberg, *op. cit.*, Vol. V, p. 100.

Chapter 1, Note # 7 Patrick Moore and Garry Hunt, *Atlas of the Solar System*. 1983, Chicago, Rand McNally, p. 220.

Chapter 1, Note # 8 Patrick Moore, *op. cit.* p. 212.

Chapter 2 - Endnotes

Chapter 2, Note # 1 Patrick Moore and Hunt, Garry, *Atlas of the Solar System*. 1983, Chicago, Rand McNally, p. 229.

Our measurements of the breadth and length of the Tharsis Bulge is 900 miles broad by 1,000 miles long. It is suspected that Moore and Hunt have an error in their estimate of the breadth of Tharsis. In any case, Tharsis is a dominating feature of the physical geography of Mars in its Serene Hemisphere.

Chapter 3 - Endnotes

Chapter 3, Note # 1 Moore, Patrick and Garry Hunt, *Atlas of the Solar System*. 1983, Chicago, Rand McNally, p. 226.

Chapter 3, Note # 2 *Loc. cit.*

Chapter 3, Note # 3 *Loc. cit.*

Chapter 3, Note # 4 *The Gilgamesh Epic and Old Testament Parallels*, translated by Alexander Heidel. 1946, Chicago, Univ. of Chicago Press, p. 87. (Tablet XI, lines 167, 170).

Chapter 3, Note # 5 Ginzberg, Louis, *op. cit.*, Vol. V. p. 240-241.

Chapter 3, Note # 6 Ginzberg, Louis, T*he Legends of the Jews*. 1938, Philadelphia, Jewish Publication Society of America, Volume V, p. 164. On the other hand, an old tradition states (Al Barceloni, 247) that each one of the seven planets has its own angel as follows; the sun has Raphael, Venus, Aniel; Mercury, Michael; the Moon, Gabriel; Saturn Kafziel; Jupiter, Zadkiel; Mars. Sammael.

Repeatedly in Talmudic material Sammael, the "angel of Mars," is reported as being an evil archangel, in contrast to the others.

Chapter 4 - Endnotes

Chapter 4, Note # 1 Hesiod, translated by Richmond Lattimore. 1959, Ann Arbor, Univ. of Michigan Press, *Theogony*, p. lines 825-829.

Chapter 4, Note # 2 Hesiod, op. cit. *The Shield of Herakles*, lines 425-427.

Chapter 4, Note # 3 Hesiod, op. cit. *Theogony*, lines 824-830.

Chapter 5 - Endnotes

Chapter 5, Note # 1 *Encyclopedia Britannica*, 1958, Vol. II, p. 574. The "iron age" in ancient thought occurred before the Noah Flood event. See Gen. 4:22. Modern archaeologists have misdated the first iron age, and have it confused with a later iron age. Ancients of the third millennium B.C. smelted a variety of metals including manganese. Perhaps they had some form of electrolysis.

Chapter 5, Note # 2 J. Lempriere, Lempriere's *Classical Dictionary*. London, 1984, Bracken Books, page 95.

Chapter 5, Note # 3 *Loc. cit.*

Chapter 5, Note # 4 Astra is that Greek goddess after which the Statue of Liberty, in New York harbor, was designed by Bartholdi. Note her tiara composed of stars. The spread of the stars in her tiara are reminiscent of an explosion in process. The Statue of Liberty was a present from the French people to the American people. Congress authorized the use of Bedloe's Island in New York harbor, already suggested by Bartholdi. The French contributed $450,000 and the American people $350,000. The statue was unveiled Oct. 28, 1886. Its height is 305.5 feet, measuring to the top of its starry tiara. Note the spread of the angles of the spikes in the celestial crown. The spread of angles resembles the spread of angles of a typical fragmentation. In New York harbor is the most visible portrait of Astra in the world.

Chapter 6 - Endnotes

Chapter 6, Note # 1 (The Sumerians had glass before the Flood, and may have cast glass into lenses and developed optical instruments. They may have seen the rings of Saturn and also Uranus).

Chapter 6, Note # 2 Moore, Patrick and Garry Hunt, *Atlas of the Solar System*. 1983, Chicago, Rand McNally, p. 241.

Chapter 6, Note # 3 Van Flandern, Tom, *Dark Matter, Missing Planets & New Comets*. 1993, Berkeley, North Atlantic Books, p. 277.

Chapter 6, Note # 4 Patten, Donald and Samuel Windsor, *The Recent Organization of the Solar System*. 1995, Seattle, Pacific Meridian Publishing. Co., pp. 69 - 108.

Chapter 6, Note # 5 Van Flandern, *op. cit.*, p. 226.

Chapter 7 - Endnotes

Chapter 7, Note # 1 Moore, Patrick and Garry Hunt, *Atlas of the Solar System*. 1983, Chicago, Rand McNally, p. 224.

Chapter 7, Note # 2 Moore, *op. cit.*, p. 219.

Chapter 7, Note # 3 Moore, *op. cit*., p. 216.

Chapter 7, Note # 4 Moore, *op. cit*., p. 419.

Chapter 7, Note # 5 Pickering, James S., 1001 *Questions Answered About Astronomy*. 1959, New York, Dodd, Mead & Co., p. 54-55.

Chapter 7, Note # 6 *Smithsonian*, August 1995, p. 71.

Chapter 7, Note # 7 Moore, *op. cit*., p. 225.

Chapter 7, Note # 8 Moore, *op. cit*., p. 224.

Chapter 7, Note # 9 Moore, *op. cit*., p. 212.

Chapter 7, Note # 10 Moore, *op. cit*., p. 225.

Chapter 7, Note # 11 *The Harper Encyclopedia of Science*. 1967, Harper & Row, p. 264.N

Chapter 7, Note # 12 Moore, *op. cit*., p. 399.

Chapter 7, Note # 13 Clube, Victor and Bill Napier, *Cosmic Serpent*. 1982, New York, Universe Books, p. 131. Clube and Napier suppose icy short term comets are originally interstellar in origin, out in the Milky Way, and entered inner Solar System space eventually to become the cause of ancient historical catastrophes.

But short term icy comets have less than one trillionth of the mass needed to create spin axis shifts and orbital perturbations of the order described herein. On the contrary, short term icy comets are the PRODUCT of Glacis during one of the planetary catastrophes, the closest to our planet. Icy comets are not the cause of ancient catastrophes, as Clube and Napier suggest, but rather are a product of one of them, which was the closest.

Chapter 7, Note # 14 Ginzberg, Louis, *The Legends of the Jews*. 1925, 1953, Philadelphia, Jewish Publication Society of America, Vol. I., p. 162.

Chapter 7, Note # 15 Heidel, Alexander, *The Gilgamesh Epic and Old Testament Parallels*. 1946, Chicago, Univ. of Chicago Press, p. 87. (Tablet 11 lines 167-173.)

Chapter 7, Note # 16 Ginzberg, *op. cit*., Vol. V, p. 182.

Chapter 7, Note # 17 Ginzberg, *op. cit*., Vol. V., p. 182-183.

Chapter 7, Note # 18 Hesiod, Translated by Richard Lattimore. 1959, Ann Arbor, Univ.

of Michigan Press, lines 194-196, p. 194.

Chapter 7, Note # 19 Hesiod, *op. cit.*, lines 141-150, p. 199-200.

Chapter 7, Note # 20 Hesiod, *op. cit.*, lines 188-196, p. 202-203.

Chapter 8 - Endnotes

Chapter 8, Note # 1 Wm. Shakespeare, *Hamlet*, Act I, Scene I, Lines 157-165.

Chapter 8, Note # 2 Few studies have been made on the behavior of composite gyroscopes, i.e. solid shell supported on laminar layers of liquids, varying in temperature, viscosity, density and gas content. Extensive behavior information for solid gyroscopes, both gimbled and not gimbled, is available. A good starting point is J. P. Den Hartog's dissertation published as a part of the *Standard Handbook for Mechanical Engineers*, edited by Theodore Baumeister (formerly Lionel S. Marks), published by McGraw-Hill, New York.

The postulated behavior of liquid core planets presented here is based upon general principles of fluid mechanics, electro-magnetic field theory and Newtonian principles of gravitational attraction. These postulations have not been tested nor have they been particularized or quantitized in representative math models in a rigorous manner.

The study of soft, planetary gyroscopes should prove rewarding as well as interesting. The effects on magma of the "stirring" resulting from spin axis precession would include changes in temperature gradients, chemical compositions, electric current flow, magnetic field perturbations, thinning or thickening of the crust, etc. This is all in addition to the geologic changes in the crust, oceanographic changes on the crust, and the extra-terrestrial electrical effects during the proximity of the two close passing planets.

Chapter 8, Note # 3 C. Warren Hunt, *Environment of Violence.* 1990, Calgary, Polar Publishing, p. 156 ff. Silanes are silicon hydrides.

This theory of the genesis of petroleum pools has two parts. The aspect involving the mixing of hydrogen with carbides to create silanes, hydrocarbons and metal hydrides, credited to C. Warren Hunt, a Canadian petroleum geologist with more than five decades of field experience. Hunt is a catastrophist, terrestrial if not yet a Solar System catastrophist.

The other aspect of this theory is the mixing mechanism. It involves sudden, nearby Mars flybys, and also requires rapid and sudden mixing of highly reactive molecules that are common to the Earth's interior, such as silanes, carbides and various other hydrides with water at the intimate molecular level, and at high temperatures and pressures. Such catastrophic mixing as a precedent for the genesis of petroleum accumulations is the concept of your authors.

Chapter 8, Note # 4 James Strong, Strong's *Exhaustive Concordance of the Bible.* 1890, Nashville, Abingdon Press, Hebrew and Chaldee Dictionary, p. 69.p.

Chapter 8, Note # 5 Raphael G. Kazmann, *On the Orientation of Ancient Temples and Other Anomalies.* Aeon II, 2, p. 51.

Chapter 8, Note # 6 Kazmann, *Loc. cit.*

Chapter 8, Note # 7 Josephus, *Antiquities of the Jews.* Vol. II, Book 9, Chap. 8, (p. 237.)

Chapter 8, Note # 8 Josephus, *Loc. cit.*

Chapter 9 - Endnotes

Chapter 9, Note # 1 For his cutting edge work, Kepler was harassed from job to job and from city to city, from Graz to Prague to Linz to Sagan to Regensburg. The harassment was by the *Medieval Association for the Advancement of Science*, Austrian branch. His works were banned by the Vatican, and were poorly received by his fellow Lutherans.

However, thankfully, the Calvinists of Strasbourg and the German Palatinate, their printers and book sellers were more receptive, objective and open-minded. From Strasbourg, Kepler's discoveries, including his Rudolphine Tables, were distributed to Geneva, Basel, Bern, Zurich, Tubingen, Amsterdam, Dordrecht, Utrecht, Edinburgh and Cambridge. At Cambridge, with his three laws of planetary motion well received, Kepler became the mentor of Edmund Halley of cometary fame, and of Isaac Newton of Principia fame.

Chapter 9, Note # 2 In the 1950's, Immanuel Velikovsky studied various ancient calendars. He concluded that the Earth formerly had a 360-day year, which meant a slightly smaller orbit, and the Moon had a 30-day orbit, which means that the Moon once had a slightly larger orbit.

His research showed that Mars had assaulted the Earth several times early in the 1st millennium B.C.E. He discovered a 600-year false expansion in the accepted dating of the Egyptian Middle Kingdom. It was an expansion serving and harmonizing Egyptian history with other gradualist expansive dating schemes.

Velikovsky also had his shortcomings. On the negative side, he had no bar charts, no graphs, no figures, no tables and just one map.

Velikovsky thought as a psychiatrist and as an ancient historian, and he was not given to mathematical, engineering or geographical analysis. So his errors and inadequacies also were big.

Despite his substantial contributions, in the 1970's, Velikovsky was pilloried and smeared by the American Association for the Advancement of Science, astronomical department. He was discredited among all branches of science. His critics were out of balance; he should have been given credit for his correct insights. But no, the scientists Donald Goldsmith, Carl Sagan, science writer Isaac Asimov, et al could not or would not "separate the baby from the bath water."

Now, in the late 1990's, your authors feel that human nature yet to improve. Treatment of this model and associated material by scientific associations, controlled more by tradition and politics than by objectivity, is expected to be no better than the treatment given to Kepler 400 years ago, and no better than the treatment awarded to Velikovsky 25 years ago.

Chapter 9, Note # 3 Edwin R. Thiele, *The Mysterious Numbers of the Hebrew Kings*. 1965, Grand Rapids, Wm. B. Eerdmans, p. 205. This was a republication of an earlier edition by the Univ. of Chicago Press.

Chapter 9, Note # 4 Louis Ginzberg, *The Legends of the Jews*. 1928, Philadelphia, The Jewish Publication Society of America, Vol. VI. 362. "On the first night of Passover as the "night of miracles", see note 76 on vol. I, p. 224. Pan Aherim 93 = Yalkut II, reads: When Rabshakah [the Assyrian commander in chief] heard the singing of the Hallel he counseled Sennacherib to withdraw from Jerusalem, as on this night - the first night of Passover - many miracles were wrought for Israel. Sennacherib however did not accept the wise counsel given him.

Chapter 9, Note # 5 Immanuel Velikovsky, *Worlds in Collision*. 1952, New York, Doubleday, pp. 330-345. He found ancient 360-day calendars among the ancient Hebrews, Egyptians, Greeks, Romans, Assyrians, Babylonians, Vedic Indians, Persians, Mayans, Pre-Incans of Peru, Chinese, Japanese and Arabians. He found traditions of 9-day weeks, 40 per year among the ancient Romans, Celts, Lithuanians, Mongols and tribes of West Africa.

Chapter 9, Note # 6 According to Table XIII, the Earth's orbit radius expanded after the Final Waltz from 361.628345 new days to 365.256365 new days. This is an increase in the Earth's period of 0.9932804%. It is about 3.628020 days.

The difference between the old spin rate, approximately 360.0 old days, and 361.628345 new days, is 0.452318%. 0.452318 multiplied by the current spin rate, 23.9345 hours (1,436.07 minutes sidereal) is .64956 of a minute. Thus it is estimated that the increase in the spin rate of the Earth, as Mars pivoted in it during the final Martian flyby, a unique outside flyby, was .65 of a minute, or 39 seconds per day.

Chapter 9, Note # 7 Immanuel Velikovsky, op. cit., p. 337. "In his Isis and Osiris, Plutarch describes by means of allegory the change in the length of the year. 'Hermes playing at draughts with the moon, won from her the seventieth part of each of her periods of illumination, and from all the winnings he composed five days, and intercalated them as an addition to the 360 days.'

Chapter 9, Note # 8 David Talbott, "Servant of the Sun God." *Aeon II*, 1, p. 37 ff. Talbott brings a comprehensive summary of the Mars archetype in world mythology.

Chapter 9, Note # 9 Ev Cochrane, On Mars and Pestilence." *Aeon III*, 4, p. 59 ff. See also Aeon III, 3, p. 71 ff, Aeon II, 4 p. 49 ff and Aeon III, 6, p. 70 ff.

Chapter 9, Note # 10 Hesiod, *The Shield of Herakles*. Translated by Richard Lattimore. 1959, Ann Arbor, Univ. of Michigan Press, p. 199-200, lines 141-150.

Chapter 9, Note # 11 Hesiod, *op. cit.*, p. 210-211. Lines 318-326.

Chapter 9, Note # 12 Ginzberg, *op. cit.* Vol. 4, p. 267-268.

Chapter 9, Note # 13 Ginzberg, *op. cit.*, Vol. 4, p. 269.

Chapter 9, Note # 14 Ginzberg, *op. cit.*, Vol. 6, p. 363.

Chapter 9, Note # 15 Patten & Windsor, two essays. "The Origin and Decay of the Earth's Geomagnetic Field" (published by Catastrophism and Ancient History), and "Clashing Magnetic Fields" (published by *Aeon*). Both are available at Pacific Meridian Publishing. Co.

Chapter 9, Note # 16 Ginzberg, *op. cit.*, Vol. 6, p. 363.

Chapter 10 - Endnotes

Chapter 10, Note # 1 Donald W. Patten, *The 108-Year Cyclicism of Ancient Catastrophes*. 1990, Seattle, Pacific Meridian Publ..F2 Robert E. D. Clark, *Science & Christianity, a Partnership*. 1972, republished 1994, Seattle, Pacific Meridian Publishing. Co., 1994, p. 53.

Chapter 10, Note # 2 Robert E. D. Clark, *Science & Christianity, a Partnership*. 1972, republished 1994, Seattle, Pacific Meridian Publishing. Co., 1994, p. 53.

Chapter 10, Note # 3 *op. cit.*, p. 41.

Chapter 10, Note # 4 Robert E. D. Clark, *Science & Christianity, a Partnership.* 1972, republished 1994, Seattle, Pacific Meridian Publishing. Co., 1994.

Chapter 10, Note # 5 Donald W. Patten and Samuel R. Windsor, *The Recent Organization of the Solar System.* 1996, Seattle, Pacific Meridian.

Chapter 11 - Endnotes

Chapter 11, Note # 1 Hesiod, *The Shield of Herakles*, translated by Richmond Lattimore. 1959, Ann Arbor, Univ. of Mich. Press. Lines 444-456.

Chapter 11, Note # 2 Neil F. Michelsen, *The American Heliocentric Ephemeris.* San Diego, ACS Publications, 1981, p. designation - March-April 1995.

Chapter 11, Note # 3 Richard L. Thompson, *Vedic Cosmography and Astronomy.*1990, Los Angeles, Bhaktivedanta Book Trust, p.91. The point we would like to make here is that the Bhgagavatam, with its 360-day year, may seem naive, but there is actually considerable sophistication behind its calculations. ... SB 5.22.15 states that Jupiter travels through one sign of the zodiac in one Paritatsara. The names Samvatsara, Parivatsara, Idavatsara, Anuvatsars, and Vatsara all refer to a year of 360 days.

Chapter 11, Note # 4 Bowditch, *The Nautical Almanac for the Year 1991.* Washington, U. S. Govt. Printing Office, pp. 10 - 250, even pages. Hamal's longitude is listed on 120 pages.

Chapter 11, Note # 5 2,699 years / 71.67° = 37.65° of luni-solar precession since the Final Mars Flyby (and no planetary precession).

Chapter 12 - Endnotes

Chapter 12, Note # 1 Vine DeLoria Jr. *Red Earth, White Lies.* 1995, New York, Scribner, p. 43. De Loria is a Sioux Indian, a leader of AISES, the American Indian Society for Engineering and Science, and a faculty member of the history dept., Univ. of Colorado, Boulder.

Chapter 12, Note # 2 Immanuel Velikovsky, *Worlds in Collision.* 1950, New York, Doubleday, pp. 330-342.

Chapter 12, Note # 3 Frank Parise, ed., *The Book of Calendars.* 1982, New York, Facts on File, p. 216.

Chapter 12, Note # 4 Parisi, *op. cit.*, p. 269.

Chapter 12, Note # 5 Parisi, *op. cit.*, p. 128.

Chapter 12, Note # 6 Jonathan Swift, *Gulliver's Travels*. New York, 1958, Random House, p. 134.

Chapter 12, Note # 7 Robert S. Richardson, *Mars*. 1964, New York: Harcourt, Brace and World, Inc., p. 93. In specifying the distance of Phobos from Mars, Swift also was specifying that to a Martian, Phobos it would rise in the west and set in the east, which Phobos (and only Phobos) does.

Chapter 12, Note # 8 Isaac Asimov, *The Kingdom of the Sun*, 1960, London, Abelard-Schuman, p. 128-129.

Chapter 12, Note # 9 Charles McDowell, "Catastrophism and Puritan Thought," *Symposium on Creation VI*. 1977, Seattle, Pacific Meridian, p. 57-60.

Chapter 12, Note # 10 Hesiod, *op. cit.*, various lines.

Chapter 12, Note # 11 *Roget's International Thesaurus*, Fourth edition. New York, Harper Collins, 1977, p. 891.10 to 891.17.

Bibliography

Abell, George, *Exploration of the Universe.* 1964, New York: Holt, Rinehart & Winston.

Alfven, Hannes and Gustaf Arrhenius. *Evolution of the Solar System.* 1976, Washington D.C., NASA.

Apollodorus. 1921, Cambridge, Harvard Univ. Press, Translator: Sir James George Frazer.

Baker, Robert H., Astronomy. 1959, New York, Van Nostrand.

Baker, Victor R., *Catastrophic Flooding: The Origin of the Channeled Scabland.* 198l, Stroudsburg, Pa., Dowden.

Bakker, Robert T., *The Dinosaur Heresies.* 1986, New York, Morrow.

Barnes, Thomas G., *Origin and Destiny of the Earth's Magnetic Field.* 1973, El Cajon Calif., Creation Research Institute.

Baumeister, Theodore and Lionel S. Marks, Editor, *Standard Handbook for Mechanical Engineers.* 1967, New York McGraw-Hill, 7th Ed.

Baxter, John and Thomas Atkins, *The Fire Came By (The Tunguska Bolide).* 1976, New York, Warner.

Bergren, W. A. and John A. Van Couvring, *Catastrophes and Earth History, The New Uniformitarianism.* 1984, Princeton: Princeton Univ. Press.

Berossus, *The Babylonaica of Berossus.* 1978, Malibu, Calif., Undina Publishing.

The Bible

Bixler, R. Russell, *Earth, Fire and Sea.* 1986, Pittsburgh, Baldwin Manor Press.

Boorstein, Daniel J., *The Discoverers.* 1983, New York: Random House.

Bowditch, Nathaniel, LLD., *American Practical Navigator.* 1966, Washington, U. S. Navy Hydrographic Office, U.S.G.P.O.

Carey, S. Warren, *Theories of the Earth and Universe.* 1988, Stanford, Stanford Univ. Press.

Caspar, Max, Kepler. 1993, New York, Dover. *Index of Subjects and Places* by Owen Gingrich and Alain Segonds.

Ceram, C. W., *Gods, Graves and Scholars.* 1951, New York, Alfred Knopf.

Clark, Robert E. D., *Science and Christianity, A Partnership.* 1971, Pacific Press, Mountain View, Calif.

Clube, Victor and Bill Napier, *The Cosmic Serpent.* 1982, New York, Universe Books.

Corliss, William R., *Ancient Man: A Handbook of Puzzling Artifacts.* 1979, Glen Arm Md., The Sourcebook Project.

Corliss, William R., *The Unknown Earth.* 1980, Glen Arm Md., The Sourcebook Project.

Crossthwaite, Hugh, *Ka, A Handbook of Mythology, Sacred Practices, Electrical Phenomena and their Linguistic Connections in the Ancient Mediterranean World.* 1992, Princeton, Metron.

Custance, Arthur C., *Without Form and Void.* 1970, Brockville, Ont., Doorway Papers.

Davis, Calvin Victor (Ed. in Chief.) *Handbook of Applied Hydraulics.* 1942, New York, McGraw-Hill.

De Grazia, Alfred, *The Burning of Troy.* 1984, Princeton, Metron.

De Grazia, Alfred, *Cosmic Heretics.* 1984, Princeton, Metron.

De Santillana, Giorgio and Hertha von Dechend, *Hamlet's Mill.* 1969, Boston, Gambit.

The Epic of Gilgamesh. 1960, London, Penguin Books, N. E. Sandars, translator.

Escobal, Pedro Ramon, *Methods of Orbit Determination.* 1965, Malabar, Fla., Robert. E. Krieger, Publishing.

Fasold, David, *The Ark of Noah.* 1988, New York, Wynwood.

Filby, Frederick A., *The Flood Reconsidered.* 1970, Grand Rapids, Zondervan.

Finch, Vernon C. and Glenn T. Trewartha, *Elements of Geography.* 1941, New York, McGraw Hill.

Force, James E., William *Whiston: Honest Newtonian.* 1985, Cambridge, Cambridge Univ. Press.

Fraser, Gordon Holmes, "The Gentile Names of God", *Symposium on Creation V.* 1975, Grand Rapids, Baker Book House.

Ginzberg, Louis, *The Legends of the Jews, Vols. I-VII.* 1937, Philadelphia, Jewish Publishing. Soc. of America.

Gleick, James, *Chaos*. 1987, Newburn, Tn., Viking Press.

Gold, Thomas, *Power From the Earth*. 1987, London, Dent.

Goldsmith, Donald et al., *Scientists Confront Velikovsky*. 1977, New York, Norton.

Gordon, Cyrus H., *Forgotten Scripts*. 1968, New York, Basic Books.

Gross, M. Grant, *Oceanography*. 1971, Englewood Cliffs, N. J., Prentiss Hall.

Guthrie, R. Dale, *Frozen Fauna of the Mammoth Steppe*. 1990, Chicago, Univ. of Chicago Press.

Hallam, Anthony, *Great Geological Controversies*. 1982, Oxford, Oxford Univ. Press.

Hartshorne, Richard, *Perspective on the Nature of Geography*. 1959, Chicago, Rand McNally.

Hatch, Ronald R., *Escape From Einstein*. 1992, Wilmington, Calif., Kneat Press.

Hawkins, Gerald S., *Beyond Stonehenge*. 1975, New York, Harper & Row.

Heidel, Alexander, *The Gilgamesh Epic*. 1949, Chicago, Univ. of Chicago Press.

Herodotus. 1929, New York: Doubleday, Ed. Gordon King.

Hesiod. 1959, Ann Arbor, Univ. of Michigan Press. Translated by Richmond Lattimore.

Homer, *The Iliad*. 1941, New York, Roslyn. Translator, Walter Black.

Howorth, Henry H., *The Mammoth and the Flood*. 1887, London, Sampson Low, Marston, Searle & Rivington.

Hunt, C. Warren, *Environment of Violence*. 1990, Calgary, Polar Publishing.

Hunt, C. Warren, Lorence G. Collins & E. A. Skobelin, *Expanding Geospheres.* 1992, Calgary, Polar Publishing.

Idso, Sherwood B., *Carbon Dioxide and Global Change: Earth in Transition.* 1989, Tempe, Ariz., IBR Press.

Josephus, Flavius, *The Antiquities of the Jews, Vols. I-VI.* 1828, Bridgeport, Conn., M. Sherman.

Kaula, William, *An Introduction to Planetary Physics.* 1968, New York, John Wiley.

Krisciunas, Kevin, *Atlas of the Universe.* 1989, New York, Mallard Press.

Kuiper, Gerard P., *The Earth as a Planet.* 1954, Chicago, Univ. of Chicago Press.

Kuiper, Gerard P., *Planets and Satellites.* 1961, Chicago, Univ. of Chicago Press.

Lempriere, J. *Lempriere's Classical Dictionary.* 1984, London, Braken Books.

Lichtenstein,, Aaron, *The Seven Laws of Noah.* 1981, New York, The Rabbi Jacob Joseph School Press.

Livius, Titus (Livy), *The Annals of the Roman People.* 1872, Philadelphia, McKay.

Luce, J. V., *Lost Atlantis.* 1969, New York, McGraw Hill.

Luckerman, Marvin A., *Catastrophism and Ancient History*, Vols. I-XV. Los Angeles, C & A H.

Luckerman, Marvin A., *Proceedings of the First and Second Seminars.* 1983, Los Angeles, C & A H.

Marinatos, Spyridon, "The Key to the Riddle of Minos", *National Geographic*, May 1972.

Mavor, James W./ Jr., *Voyage to Atlantis.* 1969, New York, Putnam.

MacCraig, Hugh, *The 200-Year Ephemeris.* 1949, Richmond, Macoy.

McDowell, Charles, "Catastrophism and Puritan Thought." *Symposium on Creation VI*, 1977, Seattle, Pacific Meridian.

Merrill, Ronald T. and Michael W. McElhinny, *The Earth's Magnetic Field.* 1983, New York, Academic Press.

Michelsen, Neil F., *The American Heliocentric Ephemeris*, 1901-2000. 1991, San Diego, ACS Publications.

Moore, Patrick and Garry Hunt, *Atlas of the Solar System*. 1982, Chicago, Rand McNally.

Moore, Patrick, *Atlas of the Universe*. 1984, New York, Crown Publishers.

Nelson, Byron, *The Deluge Story in Stone*. 1931, Minneapolis, Augsburg.

Nevins, Stuart E., "Evolution: The Ocean Says No!", *Symposium on Creation V*. 1975, Grand Rapids, Baker Book House.

Northrop, F. S. C., *Science and First Principles*. 1979, Woodbridge, Conn., Ox Bow Press.

Oates, Joan, *Babylon, (Ancient Peoples and Places)*. 1979, London, Thames & Hudson.

Ovid, *Metamorphoses*. 1958, Bloomington , Ind., Indiana Univ. Press.

Parise, Frank, *Book of Calendars*. 1981, New York, Facts on File.

Parrot, Andre, *The Flood and Noah's Ark*. 1953, New York, Philosophical Library.

Peterson, Ivars, *Newton's Clock (Chaos in the Solar System)*. 1993, New York, W. H. Freeman.

Pickering, James S., *1001 Questions Answered About Astronomy*. 1959, New York, Dodd-Mead.

Plutarch. 1959, New York, Dell Publishing.

Press, Frank and Raymond Siever, *Earth*. 1974, San Francisco, Freeman.

Price, George McCready, *Common Sense Geology*. 1946, Mountain View, Calif., Pacific Press. g

Rehwinkel, Alfred M., *The Flood*. 1951, St. Louis, Concordia.

Richardson, Robert S., *Mars*. 1964, New York: Harcourt Brace.

Rig Veda. 1981, New York, Penguin Classics, Ed. Betty Radice.

Scheidegger, Adrian E., *Principles of Geodynamics*. 1958, Berlin, Springer-Verlag.

Schweizer, Francois, "Resonant Asteroids in the Kirkwood Gaps and Statistical Explanation